Sven Gehrmann

Fire! In the Wadden Sea...

Or what the fauna of the North Sea urgently wants to tell us

Content

"And the second angel sounded, and something like a great mountain burning with fire was cast into the sea; and the third part of the sea became blood.

And the third part of the creatures which were in the sea, which had life, died, and the third part of the ships were destroyed..."

(Revelation 8, verses 8 and 9, after the Elberfeld Bible translation of 1905)

Note to this: This biblical text from the most recent book of the Bible was most likely written between the years 81-96 AD during the time of the Roman Emperor Domitian. If we look at the event described here as a gradual process, then the destruction of the ships could indicate that fishing boats will in future rot in the harbours for lack of catchable quantities of fish... Perhaps the last chapters of the New Testament should in future be interpreted much more ecologically than has been done in the past. And if the God of the Bible is indeed a loving God, then he will ultimately not punish human beings for their sacrilegious actions. No, he simply withdraws and leaves them to the consequences of their own actions... I wonder if they would then learn something from this?

There are warm times in the southern North Sea. Now, in 2020, the heat wave below sea level, which the majority of our population has never seen before, has reached a completely new quality for the first time. This is because the season known as "winter" from the days of the fathers has apparently been completely absent here since January 2020. On 10.Jan 2020, for example, the German Federal Maritime and Hydrographic Agency (BSH) reported a temperature of 10-11° Celsius off the East Frisian Islands. Although the water temperatures near the coast were somewhat lower in the single-digit range, they were still so high that it was still possible to catch small sea creatures in the tidal flats with the frame landing net, which otherwise prefer to spend the winter in deeper water zones. And also the glass shrimps of the harbour sheet piles were still present. Since January 2020, state authorities should actually have had red warning lights shining all day long. And a monotonous, muffled howl of alarm should warn the general public of the impending climate disaster. But still nothing happens. When talking to employees of national park institutions, one hears remarks like: "It is not our job to make political statements. We only exhibit something here, and the visitor can then form his own opinion." Great, I think. Because when I look at the aquaria of this "environmental exhibition", for example, I notice that although the well-groomed animals were carefully labelled. But that there's no indication of how they relate to the climate change. Moreover, the aquaria all look wonderfully tidy and clean, creating the illusion that you are looking at an ideal and intact environment. But that most of the animals shown here actually live on the seabed on a large garbage dump, which consists of PVC, PCB, net remains, car parts and other garbage, cannot be seen here. As an uncritical visitor of such an exhibition you go home and only think: "Oh great, everything is fine again in the North Sea, thanks to the National Park". Really? In this book, I'm going to point out some uncomfortable things. You'd better prepare yourself for something that may blow apart your former imaginations...

1983, Borkum: I, then 14 years old, got the chance of a lifetime: I was allowed to go out with a real professional fisherman. To catch shrimps! I remember how we lowered the net on 07.07.1983 at clear sight off the bird island Rottum. Two mighty beam trawls dragged evenly over the ground on each side of the ship. After an agonizing three quarters of an hour, the mighty beam trawls were caught by a winch. Full of joyful anticipation I hopped over the deck and - much to the annoyance of the fisherman - almost got the beam trawlers on my head. The net was full of true shrimps, also known and dealt as "crabs". Great pipefish and tub gurnards particularly fascinated me at that time. We also caught lots of flatfish of all sizes, various eels and soles, the latter two of which we boiled fresh on board. I have never eaten better fish! And today?

2003, Baltrum: A short holiday with the family. Recently, giant oysters have been appearing here in the waddens; occasionally on stones. It's April, the sun shines so much that the islanders have to turn on their sprinklers, because the grass on the island starts to wither. Furthermore, I find **Pennant`s Swimming Crabs (*Portumnus latipes*)**, which had their origin off north-west African coasts, washed up on the beach. All females, that came to the southern North Sea to reproduce here...

2011, Norddeich: In the harbour basin small fishes swim on the surface, just about two centimetres long. An investigation shows that these are juvenile **Sea Bass (*Dicentrarchus labrax*)**. Not a single flatfish can be caught with the landing net in the local tidal flats... The pier is covered with **Pacific Giant Oysters (*Magallana gigas*)**. The once common **Blue Mussels (*Mytilus edulis*)** have become a scarce commodity here...

2012, Baltrum: It's midsummer in August. At high tide, anglers are still standing on the groynes. What they do catch here? Sea bass; the island record is something about 70 centimetres long...

2012, Norddeich: This time no sea bass in the harbour basin, but small flatfish in the waddens... At least; but only a few.

2013, Norddeich: With the bait sink net only little animals can be detected in the harbour basin. But also an imported shrimp from Korea, the **Oriental Glass Shrimp, *Palaemon macrodactylus.***

2014, Norddeich: And again in April a shrimp trawler brings along egg-bearing females of the subtropical **Arch Fronted Swimming Crab *Liocarcinus navigator***. The North Sea is too warm for the season. Summer?

Spring 2015 and 2016, Norddeich: The shrimp trawlers catch tope sharks, broadnosed skates, anchovies... All immigrants from the English Channel. The Winter 2014/2015 was once again much too warm for our latitudes...

2017, Filthy weather in East Frisia: Not really a summer, it's constantly muggy or rainy, the farmers have many problems to be able to harvest anything at all... The by-catches of the fishermen are very different, certain otherwise common species are rare...

2018: Heat wave! Many otherwise common fish species were hardly caught by fishermen in summer. Because with a water temperature of 22° Celsius in the southern North Sea, they prefer to stay in deeper areas where no trawlers fish... In the Baltic Sea: 25° Celsius and vibrio alarm! In addition to this, one could observe more jellyfish than usual. Have they decimated the fish brood?

2019: Almost the same as 2017, only considerably warmer. In summer, the tub gurnards and other otherwise frequent by-catches are missing...

On 31.Dec 2019 it was still possible to catch true shrimps in the tidal flats of Neßmersiel. The North Sea is too warm. Winter? What is winter?

2020: January. Temperatures up to the double-digit range. Spring bloomers already have buds. There are already daisies... The season "winter" seems to be gone this year in East Frisia, Lower Saxony. The shrimp fishermen go out, although they usually take a winter break at this time of the year. And they bring unusual creatures into the harbour, which is evidence of unbelievable things. As we will see in the following... Some of these catches are inconspicuous and small, others are almost screamingly colourful. As if Mother Nature wanted to send us a brightly coloured warning before it's too late...

1. Temperatures in the North Sea (i.e. the German Bight)

According to the Federal Maritime and Hydrographic Agency (BSH), the average water temperature of the North Sea in 2017 reached 10.9 degrees Celsius, only slightly below the 2016 value of 11.0 degrees. This was the second highest value since 1969, and only in 2014 was the water even warmer at 11.5 degrees Celsius. At the moment (January 2020) we have temperatures between 10 and 11 degrees Celsius in the southern North Sea off the East Frisian Islands... You can catch true shrimps in the waddens, which should actually be hibernating in deeper water layers... The "winter" season does not seem to be taking place in the southern North Sea at all at the moment...

2. On 18.01.2020 at the internet provider GMX it could be read that...

...a heat wave in the ocean caused a mass death off the American coast of the USA. Researchers from the United States called the stream of excessively warm seawater that stretched from Alaska to the Baja California a "blob". This caused the death of several ten thousand of birds in the Pacific, but also millions of starfish. The warming of the sea water favoured the reproduction of viruses and epidemics, which are responsible for the death of these animals. Climate change could make such deadly heat waves in the sea more and more frequent. Incidentally, the seabirds simply starved to death because the warm seawater caused their otherwise abundant prey to disappear... Scientists estimated the total number of dead seabirds at about one million... They assumed that the "sea heat wave" had reduced the quantity and quality of the plankton, so that the number of fish living on it was greatly reduced. In addition, the metabolism of fish in warmer water was said to have become faster. As a result, the predatory fish would have needed much more prey than usual due to the resulting higher energy conversion, thus further reducing the number of available prey fish for seabirds. In addition, other creatures had also been affected by the problem of ocean warming. Among others, "only" about 100 million cods have died. And many whales also suffered from the consequences of this dramatic ocean warming.

3.) Sea heat waves caused by a climate warming up

These already existed in the Tasman Sea and in other regions, such as the Australian Barrier Reef or on the coasts of South America, where the phenomenon has been known for many years under the name *El-Niño**.
In 2018, these phenomena could also be observed very clearly in the German North Sea and in the southern Baltic Sea... In summer 2018, the North Sea became 22° Celsius warm, while the southern Baltic Sea even reached temperatures of up to 25° Celsius. These were values as they are usually measured in the Mediterranean Sea...

4.) The warming up of the oceans is accelerating more and more...

Scientists have calculated that in 2019, the world's oceans will be warmer than at any time since the global survey began. They also warned in the journal "Advances in Atmospheric Sciences" that climate change is accelerating the warming up of the oceans. According to the report, the past ten years have seen the highest ocean temperatures since the 1950s, with the last five years also being the warmest. In 2019, the average sea temperature down to a depth of two kilometres would be about 0.075° Celsius above the average from 1981 to 2010.

5.) A preliminary review...

This enormous amount of energy in the form of heat emitted into the oceans by man over the past 25 years is equivalent to the energy discharge of 3.6 billion(!) atomic bomb explosions of the size of the atomic bomb, that was detonated over the Japanese city of Hiroshima during World War II.

*Spanish = the Christ Child. The phenomenon of the warming of the Pacific Ocean off the coast of Peru from 24° to 26° Celsius has been known for many years and occurs sporadically about every 4 to 5 years, but has increased in intensity significantly in recent years. This warming up destroys first the plankton and after that complete fish stocks...

This book, of course, makes no claim to complete scientific accuracy or a fact-oriented, all-encompassing truth. Rather, the author has attempted to assemble a picture from the very confusing puzzle pieces and fragments of shrimp trawler catches, his own observations and the messages of third parties. This picture is of course neither complete nor completely finished, because the puzzle pieces entrusted to the author are simply too few for this. But even from the few pieces of the puzzle, valuable small pieces and aspects of a larger whole can already be recognized.

The author's horizon of observation extends, with interruptions, from the late 1970s and early 1980s until today.

Although these observations are largely based on a certain amount of subjectivity, they become more objective because it has been possible to shed some light on certain events through personal experiences and messages from third parties.

For example, some personal communications from the aquarium operators on the island of Borkum compensate somewhat for the "absences" of the author on this island, during which he could not be present there.

The author's collection, with which the discovery of certain animal species could be assigned to specific periods of time, also proved to be helpful in the creation of this work. Thus, for most of the hypotheses described here, there are real specimens that can be checked at any time. And so, in the meantime, about 500-600 jars with corresponding specimens have accumulated in the author's cellar. It should also be noted that the animals preserved here were not collected and killed just for collection purposes, as most scientists would do.

But more than 95% of these specimens are animals that were either collected dead between garbage and by-catch, or that were later placed in an aquarium. So there was no overexploitation of the endangered fauna of the North Sea, just to extend a collection. But even if we were to look closely at all these collection specimens, even these would only represent a very small part of the creatures that actually exist in the southern North Sea. Because it is estimated that in the whole North Sea there are at least 7000 different species of animals, of which about 4000 live in the German Wadden Sea. And there are probably even considerably more species if one were to take into account species introduced from other continents, and if one were to include the world of microscopic marine organisms, which is left out here in this work. In order to draw conclusions about climate change from marine organisms, several aspects must be considered in context.

For example, it is a very important premise to first determine which species originally dominated the habitats of the southern North Sea. Then a careful comparison over time can be made and finally, in the final step, new species can be identified. The respective sizes of the specimens found and the new frequencies of these species play a very important role.

And if one looks at different groups of animals, a quite objective pattern emerges, which clearly shows the warming processes in the southern North Sea along the coast and off the East Frisian Islands.

Above all, it is a question of an overall trend that clearly shows that the deniers of anthropogenic climate change must either be completely blind or deaf. Or even suffer from the worst failures of human nature: namely, stupidity, ignorance, consumerism, or selfishness. This book should help to show how far climate change has already come. And that it is by no means the crazy idea of left-wing eco-socialists or other environmental fanatics. The truth can be denied, but its consequences will always catch up with its deniers and opponents. Some sooner, some later. In the end, the uncomfortable truth in this small tributary sea of the Atlantic will take its toll on all of us, whether we like it or not. The little colourful animals that have recently been romping in our waters should be better understood by us as a reminder from Mother Nature to all of our consumer behaviour. Because only if our philosophy of life and basic attitude change, we still have a future worth mentioning on this coast, which is still valuable for tourism...

Autochthonous animal and plant species are generally defined as those species that have been known for long periods of time from specific localities. Basically the term autochthonous means as much as native or long-established. Not all autochthonous species are threatened by climate change, because there are actually some species that benefit from a warming up of the North Sea. The benefit then lies in opening up new ecological niches through the disappearance of other species, better reproductive opportunities or improved food supply. The latter could, for example, be the result of a more abundant growth of phytoplankton or other marine algae as a result of longer growth periods. Because the warmer the ocean is, the faster the tiny planktonic algae can multiply through cell division. However, there are (still) a number of animal species living in the German Bight which belong to the boreal or subarctic faunal group. And these will either migrate into deeper, colder areas on the sea floor or towards the north. At a certain point of no return, they will simply stay completely away from the shallow waters of the German Wadden Sea, especially the southern North Sea. Now, this may not have too drastic an impact at first, if only a few subarctic species such as the great spider crab or the hooknose get lost from the ecosystem of the German Wadden Sea. Other species with a higher temperature tolerance will certainly move in from the south and try replace them. But in the medium and long term, such creeping changes could very well cause major and dramatic collapses, for example in fisheries or in the feeding of migratory birds. A lack of faunal components automatically results in a different composition of the marine plankton, which could be decisive for the later number of useful fish in the sea and of nutrients for migratory birds in the tidal flats. This is because ecosystems are highly complex structures. The fate of herring and lobster is therefore already decided in the plankton, which can react very sensitively to changes in its environment, even at short notice...

A further problem of autochthonous species of the boreal climate zone is that they are adapted to very specific temperature ranges for the completion of their life cycles. In the North Sea, these are linked to specific localities, currents and seasons. If these parameters shift, the species affected may lose food sources or habitats. Or, in the worst case, the possibilities for reproduction of their species, which then leads to the extinction of the species. For example, some species are very dependent on certain sea temperatures, because oxygen levels and the thriving of broods depend on them. In March 2017, for example, there was a massive slump in smelt fishing in the Elbe estuary, and even in spring 2020 things looked bad for these once very common small coastal fish. In addition to the problem of water warming due to climate change, there was also sediment turbulence as a result of dredging and the extraction of cooling water for power plants, as well as the discharge of warm power plant waste water. It is also known from many fish species of the cold temperate faunal cycle that they require cold periods in winter to mature their gonads. If these are too short or do not take place at all, the species affected by them become infertile. In addition, the phenomenon that a warming up of the environment surrounding an organism accelerates its metabolism. This means that its metabolic processes now run faster and faster. The heart races, breathing becomes more hectic and the need for food increases rapidly. As a result, our autochthonous species can only adapt to an ever faster increasing climate change to a very limited extent, because either they collapse themselves or they are no longer able to produce healthy and viable offspring. It is already a de facto situation, that commercial fishing fleets have to travel more than 100 kilometres further north than before, to be able to fish significant quantities of cod and saithe. As these have already begun to migrate towards Norway and Iceland. In 2009, the cod stocks in the North- and Baltic Sea had already collapsed. And the same happened in 2009 in the southern North Sea with flatfish. At present (2020), plaice and sole are considered to be rather scarce in the southern North Sea...

Since its foundation in 1892, the Biological Institute on Heligoland has been measuring and documenting the water temperatures in the North Sea at different times of the year, from which an annual average value has always been obtained. The result of these measurements is clear: since the last 50 years alone, the North Sea has warmed up by about 1.7° Celsius. And since the beginning of these measurements more than 100 years ago, this value has probably even risen by 2° - 3° Celsius. Now one could argue that this is only a very small rise in temperature, hardly perceptible for a person bathing. But: The world ocean is on average only about 3.8° Celsius warm anyway, if you include the water masses of the deep-sea trenches and the Arctic and Antarctic regions. If one were to relate these values, i.e. 3.8° world ocean average and 1.7° North Sea warming by simple rule of three calculation, the North Sea would have warmed up by 44.73% in the last 50 years compared to the entire world ocean. If we disregard this approach and only put the warming of the North Sea from a temperature of about 9.5° Celsius in 1965 in relation to a temperature increase of 1.7° by the year 2015, then the North Sea has still warmed up by almost 18% in only 50 years! The rise in temperature in the North Sea (the German Bight) is thus an officially proven fact and seems to be developing disproportionately to the other global warming, as this was only about 1° Celsius in the same period. This is mainly due to the fact that the North Sea as a whole is a relatively shallow shelf sea with an average depth of only about 70 metres, which in addition has only one deep basin in the north, the Norwegian basin, which is about 700 metres deep. In addition, there are also very extensive waddens, which regularly dry out very extensively due to the tidal range and heat up particularly strongly in summer thanks to the sunlight on these dark areas. Thus, the animals and plants of this tidal habitat have to cope with various extreme life situations. In winter with icy cold, but in summer with constantly increasing subtropical and tropical temperatures. In some cases they adapt to these changes at short notice, in other cases they migrate or simply die out. An example of a temporary adaptation to higher sea temperatures is the large brown seaweed on the island Heligoland, which, according to the observations of marine biologists, has started to anchor its rhizomes at ever greater depths on the rocks in front of the island. They do this not, because they have developed a lower demand for sunlight, but rather, because the surface water has simply become "too warm" for them in the summer months. Seaweeds react very sensitively to warmer water, which you can test for yourself in a cooled marine aquarium. If the water becomes too warm - and if it is even a single degree Celsius higher - the kelp may suddenly become slimy and dissolve. Some seaweeds then release gametes through this self-abandonment, through which they reproduce in order to at least ensure the survival of the species elsewhere. In nature a sensible adaptation to changing living conditions - in the aquarium usually a catastrophe for all other inhabitants of the tank, who can be downright poisoned by this. But also in nature a mass death of algae or seaweeds can lead locally to a mass death of fish and other marine animals. Especially in shallow sea bays and waddens, which can be covered by algae in a short time. This may then be accompanied by the development of harmful bacteria and viruses, which not only cause "red tides" and fish mortality, but can also become dangerous to humans themselves. Like the thermophilic vibrios. These germs can even enter the human organism through small open wounds, where they can cause severe diseases, especially to immunocompromised persons, such as sepsis and in consequence sometimes even death! Probably the best known of these pathogens is the **Cholera bacterium *Vibrio cholerae***, which causes cholera. The **Wound bacterium *Vibrio vulnificus*** occasionally makes headlines by causing cellulite and sepsis in immunocompromised persons, especially in the shore zone of the Baltic Sea during the summer heat, sometimes with a life-threatening course. So a sea warming up in summer in shallow water is anything but a harmless natural phenomenon! In the following chapters we will look at some very different groups of invertebrates which have now more or less become native to the southern North Sea off the East Frisian Islands. Some of them used to be "seasonal animals", which could be found sporadically only for a few days or weeks in the German Bight. But today they have become part of the "permanent population" of this area. Two things prove this: First the establishment of local populations, which suddenly appear regularly in certain areas. And second the presence of their juveniles. Which suggests, that they not only spend the winter here, but can also reproduce. So what was the rare exception in 1960 is now beginning to become the rule. This would not change much even if we were to have an extremely cold winter with arctic temperatures in 2020. It is true that the northernmost populations of immigrants would then die off in the short term. But because of the overall warm trend of the last decades, they would be back by the next warm summer at the latest and would start their populations again. However, there are several different types of immigrants, because not all species are here because of the warmer climate. Some simply came from overseas as larvae in the ballast water of ships into our waters, and can establish themselves here because of similar living conditions. Others are part of the typical English Channel fauna, that can be found on the French and British coasts. These species are not without problems in that they can either displace our native species or even mate with them, thereby destroying their genetic potential. So it is very worrying when we recently see specimens of species that have the characteristics of various related species that are difficult to identify. These phenomena can now be regularly observed in blue mussels of the genus *Mytilus* and shore crabs of the genus *Carcinus*. And then there are the species introduced by aquacultures, such as the **Pacific Giant Oyster** *Magallana gigas*, which benefit greatly from the warming up of the oceans due to their warm regions of origin in Asia. And some species have been found as single specimens, which can be regarded as vanguard for further biological input into the German Wadden Sea. I found many of these species on plastic garbage and other human relics that our shrimp fishermen had collected with their nets from the tidal flats of our German coasts. However, it is very pointless to discuss whether these garbage finds were transported from the coasts of the English Channel by the flow to the German coasts, or whether these organisms entered our waters as larvae and only then colonized the garbage. Fact is, that they suddenly appeared out of nowhere. And at least nobody had really noticed them before. In January 2020 alone, I found three(!) new species from the English Channel on a single plastic box, which could be determined quite well with the help of a Kosmos-nature book and a subsequent Internet comparison. In the following, I will go into more detail about various invertebrates, whereby I was able to relate their appearance to a certain extent to current environmental events.

Phylum *Mollusca* - Molluscs

In the North Sea you will unfortunately not find as many species of molluscs as in the tropics, because the biological diversity is lower. But there are always representatives of the Arctic faunal circle, such as the Icelandic clam, which in the deeper areas of the southern North Sea meet with immigrant species from the English Channel. Such as the common otter mussel. In the past, one faunal circle dominated slightly, then the other again. Today the Arctic faunal circle withdraws to the north, so that the Icelandic clam, for example, may soon have completely disappeared from the southern North Sea. This is because species of the Arctic faunal circle can only escape the heat to the deep or to the north. Such species then die out unnoticed at first. This means that serious conclusions must be drawn from the loss of species. No statement at all - i.e. no finding - is sometimes also a statement! In addition, the size of specimens of new species found is also important, as these can provide information about the food supply and the altered growth intervals. Especially in the case of bivalve molluscs, heat and cold phases can be read off excellently from the formed annual rings of their shells. By the way, this also allows to determine the year in which a mussel was formed and how old it actually became. This type of observation then enables the experts to draw precise conclusions about the climate development.

The **Slipper Limpet** does not belong to the endemic snails of the North Sea, as it was introduced by ships only a few decades ago. Slipper snails owe their name to the oval shape of their shell, which again contains a white inner pocket made of lime, behind which the mollusc attaches itself to its shell. Thus the empty shell of this snail is indeed reminiscent of a slipper with a cavity for the foot. Inside, the shell shines red-brownish pearly. Slipper snails live as filter-feeders, whereby they filter the finest particles out of seawater and then use them as food. But they also seem to be able to graze algae films off their substrate when necessary. However, slipper snails very rarely move on their substrate. They are also not able to turn around on their own when they are torn off the substrate. Therefore they suck themselves very firmly on their respective substrate and can only be torn off with brute force. Mostly they cling to stones, wooden posts or other shells. Slipper snails are hermaphrodites, which change their sex in the course of their life. That is why they often sit on top of each other, so that they can release their different sexual products into the water at the same time. In appropriate aquaria without significant predators, slipper snails can survive up to one year, but unfortunately, they usually starve to death due to a lack of suitable fine food. Recently some specimens could be found in plastic waste and old boxes, which had developed unusual shell shapes. Some of them were very flat and because of this shape reminded more of the slipper snail of the Mediterranean Sea. However, on the coasts of North and Central America there are also some species with a very flat shell, such as the species ***Crepidula excavata***, for example. Another specimen even had a strangely wavy shell and showed similarities with several other species of the American faunal circle. It is also conceivable, that these are only newly formed morphs of ***Crepidula fornicata***, which have now adapted to a warmer and dirtier environment...

Slipper snails often sit on top of each other for synchronous spawning.

Left and centre: These strange flat and grooved slipper snails appeared in late 2019 and early 2020 in the by-catch of a shrimp trawler attached to plastic waste. Either they are newly introduced species from the American coasts, or an adaptation to new environmental conditions.

The **Scaly Saddle Oyster** attaches itself to stones, other molluscs and to oyster shells, whereby it - like the oysters of the ***Ostreidae*** - always grows together with the lower half of the shell with the substrate. This is why this filigree mussel does not survive if you try to scrape it off the substrate, as this usually destroys the lower shell. The scaly saddle oyster has only a very thin layer of luster in its shells, and these are always very filigree and almost transparent. In January 2020 I found some specimens of this species on a plastic box, where some of them had grown exactly into the rectangular angles of the plastic grid. On the same box there were also two species of barnacles, that were not yet known from the German Bight. At first sight, such finds are very small and not spectacular, but recently a clear tendency to increase such finds has become apparent. It is unclear whether the species is currently only drifting to us in the summer season or whether it has already begun to establish itself in the southern North Sea. I had already made similar finds in 2014, but unfortunately, I did not examine them more closely, which is why it is now difficult to make a statement about the exact species affiliation of the saddle oyster species found at that time.

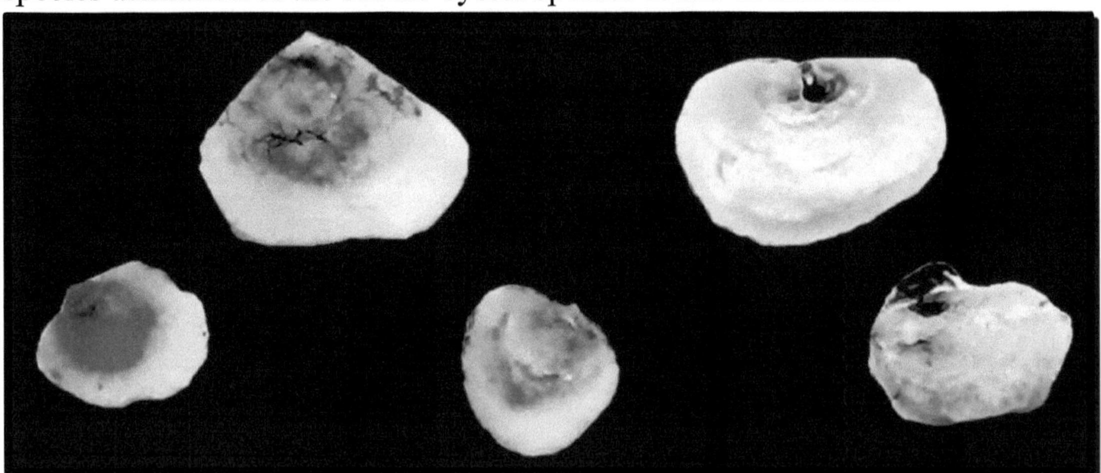

Scaly saddle oyster, *Pododesmus squama*, 40mm. Regularly found on the British coast up to Norway.

Icelandic Cyprine, *Arctica islandica* (Linnaeus, 1767)

The **Icelandic Cyprine** reaches a shell length of up to 100 millimetres. The species can be kept in a refrigerated aquarium for up to one year. In 2006 British researchers found a specimen off Iceland which, after counting the annual rings, could be dated to an age of 507(!) years. The specimen was named "Ming", because it must have started its life as a mussel on the seabed during the Chinese Ming Dynasty. This makes this Icelandic clam the most long-lived non colony-forming creature found on the planet to date. (And if Ming had not been found, the specimen might have become much older!) The shells of the Icelandic cyprine could occasionally still be found on the beaches of the East Frisian Islands until the beginning of the 2000s after storm surges. And sometimes the shells were collected by shrimp-trawlers which fished in a depth of about 20 metres and more. But even these catches and finds now seem to belong to history in the area of the southern North Sea. Also the catches or finds of juveniles, which would indicate a reproductive population of this arctic species, are unknown to me from the southern North Sea. Therefore, it is reasonable to assume, that the Icelandic cyprine is already almost extinct in the southern parts of the North Sea. It may be that a few Methuselahs of this species may still be found in isolated deep places on the sea beds, that are inaccessible for fishing. However, increasing ocean warming up has probably already de facto killed this Arctic mussel in the southern North Sea.

Pacific Giant Oyster, *Magallana gigas* (Thunberg, 1793)

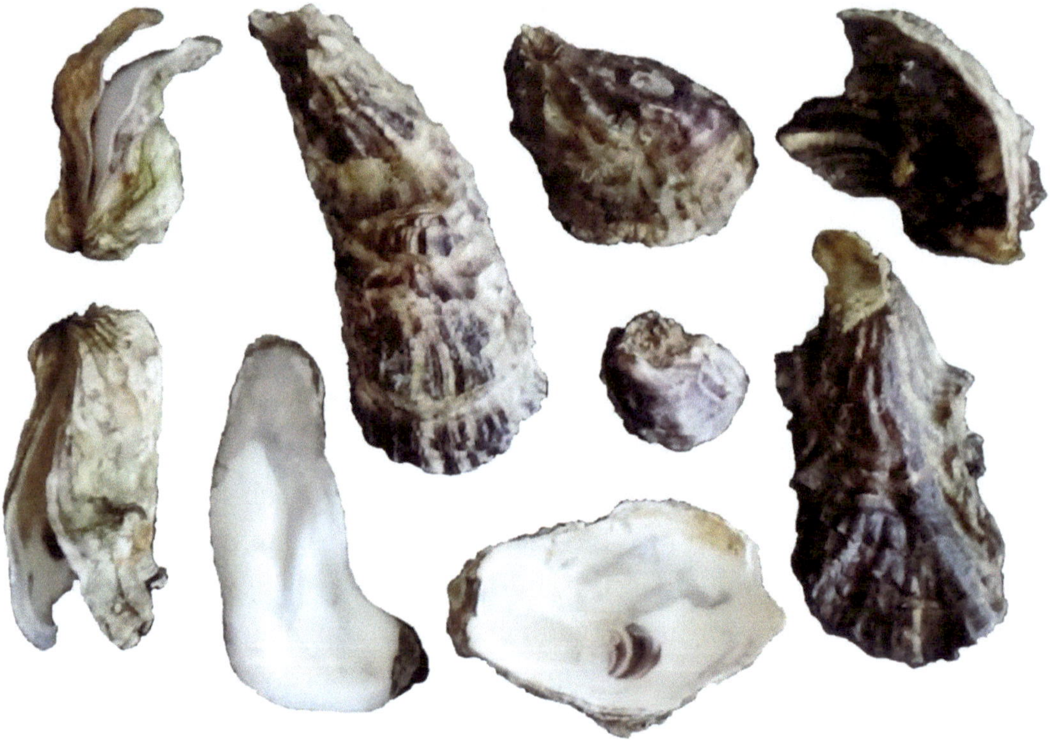

The shell shapes and colours of the Pacific giant oyster are very variable.

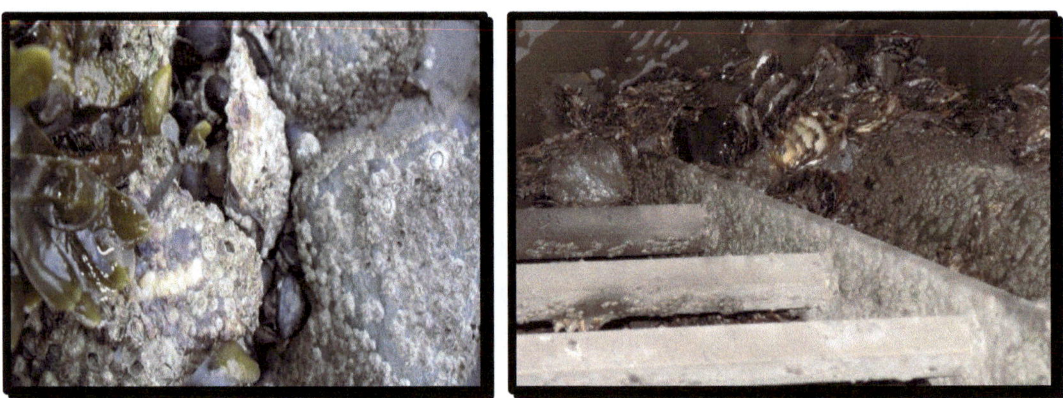

These oysters grow on stones and encrust whole sheet piles.

Pacific oysters have been found very frequently on groynes and sheet piles in the southern North Sea since the early 2000s. Sometimes they encrust complete sheet piles and stones. Because stocks of the European oyster declined dramatically as a result of marine pollution, oyster farmers imported the up to 40-centimetre long **Pacific Giant Oyster**. As a result, this species can now be found regularly in the Wadden Sea. In April 2003 I was able to discover some single specimen on the island Baltrum. Only three years later, the oysters had reproduced so strongly that I was able to find large specimens of 10 centimetres in length in large quantities on the sheet pile walls of the port Neßmersiel. In addition, small specimens up to about 7 centimetres were found on tufts of seaweed in the waddens. In 2006 and 2007, large oysters of 10 centimetres and more in length were found in the tidal flats, already torn away by the flow. Attached to them numerous other organisms were found, such as chitons, barnacles, sea anemones and slipper snails. The oysters offered an excellent substrate for snails in particular, as thin films of algae often grew on them. The ecological dimension of this oyster invasion is currently unclear. For example, the oysters could change the depth of visibility of the water due to their enormous filtering capacity. In previously cloudy places, this could mean, that seabirds could catch more small fish than before, which could lead to a decline in these stocks. On the other hand, oysters themselves offer settlement space and protection to other organisms, as their shells are very resistant and can withstand most shellfish eaters for a long time. Perhaps this is the reason why oysters have displaced the blue mussel in many harbours. The situation will become alarming, when the oysters overgrow the mussel beds and the mussel felts erode. In this case a not insignificant natural coastal protection would be lost, which could have dramatic consequences for the coastal population in the event of a storm surge. The strong reproduction of the Pacific giant oyster also indicates, that this species benefits greatly from climate change. And even short-term losses of juvenile mussels due to a harsh winter are apparently easily compensated for by this species, as could be observed in 2012 and 2013.

Queen Scallop, *Aequipecten opercularis* (Linnaeus, 1758)

Queen scallop, *Aequipecten opercularis;* right: *A. opercularis radiata*.

Variegated scallop, *Mimachlamys varia*, 50mm. This rare species also appeared in bycatch in 2018 and 2019.

The **Queen Scallop** is found on soft grounds from 15 metres depth, which is why its shells are very rarely found in the rinse of the beaches. It grows to a maximum of about 8 centimetres in diameter. These large specimens are also fished commercially in the English Channel and the Irish Sea. We do not have specimens of this size here, but it has been observed on the by-catches of the shrimp trawlers that the catches have been increasing since 2012 and have now also become somewhat larger. In 2012 I noticed for the first-time juvenile mussels of this species, which had attached themselves to a plastic foil. They were about 5 millimetres to one centimetre in size. In the following years these mussels appeared occasionally as single specimens of about 3 to 4 centimetres in the bycatch. But after the record heat of 2018, in autumn and winter 2019, several hundred specimens, including several very colourful bivalves, were suddenly collected by the shrimp trawlers off the island of Juist. These were then mostly about 3 to 4 centimetres wide. This leads to the assumption that these mussels come to us for two reasons. Firstly, they find a better foothold on the many rubbish on the sea floor than on bare sandy soils. And secondly, like almost all bivalve molluscs from the English Channel, they benefit from the ever-warmer seawater, since under such conditions the small plankton organisms of the phytoplankton, their main food source, can reproduce even better and faster.

In 2019 the fishermen of Norddeich suddenly landed hundreds of these mussels off the island of Juist. Most of them were 3 to 4 centimetres in size and there were several very colourful specimens among them, whose colours reminded of tropical scallops.

In addition, there were also a few specimens of the **Variegated Scallop, *Mimachlamys varia***. This species has been very rare in our waters so far, but it is somewhat more common in more southern parts of the sea. Should it become more frequent and larger in the southern North Sea in the future, this would be a further evidence of anthropogenic climate change.

The **Galician Mussel** can already be assigned to the southern boreal faunal circle. It is regularly found from the Mediterranean Sea unto the Dutch coast, but for some years now it appeared on the East Frisian islands and in the southern German Bight. It reaches a length of about 80 millimetres. It can be distinguished from the common blue mussel by the slightly wider and flatter basic shape of its shell, the rather brownish flesh and the slightly purple edge of its mantle. The problem is that this species can also bastardize with the **Common Blue Mussel *Mytilus edulis***, which leads to almost indeterminable specimens. The Galician mussel has been on the advance northwards in the southern North Sea since the early 2000s, while the common mussel is on the retreat. In 2013, mussel fishermen complained of a significant decrease in their yields in the German Wadden Sea in the southern North Sea. I could already find this species on Borkum, on Norderney, on Baltrum, in Norddeich and in Wilhelmshaven. However, these specimens were mostly small to medium-sized mussels up to 40 millimetres. However, it cannot be excluded that among these specimens there are also hybrids of ***Mytilus edulis*** and ***Mytilus galloprovincialis***. These small mussels are now quite often found on plastic waste, old fishing nets and other waste to which they are firmly attached with their byssus threads. Since about the middle of the last decade their frequency has increased significantly. What consequences this small invasion will have for the ecosystem is not yet foreseeable.

The **Sand Gaper** or **Steamer Clam** lives deeply buried in the waddens. It stretches out its two-part siphon, through which it swirls in its breathing water and small food particles, and the filtered water next to it flows out again. Clams can penetrate several decimetres deep into the substrate. In doing so, they penetrate into oxygen-free soil layers in which other animals can no longer live. Clams can be dug up in the waddens, and the frightened clams often spit out water fountains. They are edible, but they are only used locally by humans. They react sensitively to water pollution and coldness. Local populations can suddenly die off over large areas in the event of unfavourable environmental factors. Then the stinking dead mussels are washed up on the beach in large quantities, and even years later the accumulation of their shells bears witness to the disaster. Because of their size, clams contribute a large part to the mussel shell accumulations off the coast. For this reason, these mussel beds were even commercially exploited in the past by the coastal dwellers in order to extract lime from the mussel shells for building houses. The sand clam is distributed throughout the northern hemisphere, and in the North Sea it had temporarily disappeared during some cold periods in the past. This species is therefore benefiting from ocean warming and it is very likely that its population could grow a little more than before.

Live Common otter shell from a shrimp trawler, caught in 2017.

The shells of the Common otter shell have a completely different denotation than those of the Sand Gaper. In addition, they are often covered by a thin brown skin, also called *periostracum*. Thus, when fresh, they are more reminiscent of a freshwater pond mussel.

The **Common Otter Shell** actually belongs rather in the English Channel and has been found only sporadically on German coasts. However, due to the last warm years, a population seems to have established itself in the southern North Sea off the islands of Borkum to Norderney and Baltrum since about 2010. At first sight, this mussel resembles the **Sand Gaper** *Mya arenaria*, but its shell flaps are somewhat rounder and its shell thickness is less. And also its siphon is comparatively shorter, thicker and not quite retractable. Their shell is covered by a brown skin, which at first sight resembles a freshwater pond shell. The otter shell can grow to about 13 centimetres long. Although it does not belong to the sand gaper family, it has a similar way of life and can also dig up to 40 centimetres deep into the substrate. It is not yet possible to say whether the adder bivalve mollusc is just an example of species transfer due to shipping traffic or another faunal evidence of global warming up in the southern North Sea. This is mainly due to the fact that otter shells are among the species, which migrate northwards during short warm periods and then die during a severe winter. And thus disappear again from the habitats that have just been developed. However, it is noticeable that since the warm year 2016, shrimp trawlers have occasionally landed these mussels. In the following years 2017, 2018, 2019 and 2020, the shells of otter shells have also been found more frequently on the beaches of the East Frisian Islands. It is therefore very likely, that the fauna of the English Channel has already begun, to shift northwards. The only thing that is still missing, is the evidence of small juvenile otter shells, which would clearly indicate a reproduction of their population. However, the empty shells of juvenile otter shells are also rarely found than the similar clams. An analysis and cautious estimate of the molluscs discovered so far, which had a size of about 100 millimetres, showed, that these specimens must be about 8 to 10 years old. This would therefore mean, that the larvae of this species must have settled on the soils off the East Frisian Islands from about 2010 onwards. Their growth documents, that in between they are very well adapted to the new living conditions in the north.

Adult Forbes's Squid, about one foot long.

Forbes's Squid eggs, in Germany mentioned as "asparagus jellyfish".

The Forbes's Squid is a species, which was originally known from the Northeast Atlantic and the Mediterranean Sea, but it is now much more abundant. This means that it has extended its distribution area in the south via the Suez Canal and the Red Sea to the Indian Ocean along the African coast. On the other side, however, it is now advancing further and further northwards. Since the middle of the 2000s, it has even been known from Swedish waters. It is one of the most common species of the genus in its area of distribution and is seasonally very often caught commercially. But then not in the North Sea, but in further southern areas. With a total length of about 60 centimetres, this squid remains considerably smaller than the **Sagittate Squid**. However, the best way to distinguish it from the sagittate squid is to look at its eyes, as it cannot close them, and the shape of its fins at the end of its body. Squids normally swim forward with their fin hems and use the recoil nozzle only to escape from enemies. They can confuse the attacker at the same time as ejecting a dark liquid ink. They grab their prey with their two strongly elongated tentacles, then lead it to their parrot beak and bite it into small pieces. For several years now, Forbes' squids have apparently been spawning more and more frequently and seasonally earlier in the German Wadden Sea. Here the females lay their egg capsules in a ring around a central point like the rays of the sun. These clutches are also popularly known as "asparagus jellyfish". In the heat year 2016, these squids appeared in May and June in the southern North Sea near the East Frisian islands of Juist and Norderney, where several dozens of them were caught by shrimp trawlers while spawning. Some egg clutches were also caught, as the animals had attached them to old netting and other waste. After the "winter" season had de facto ceased to exist in the southern North Sea in spring 2020, the shrimp trawlers caught single Forbes' Squids as early as April. The catches continued until May. This proves that the animals now come to the southern North Sea to spawn earlier than usual. In the heat year 2018 they could be caught by Büsum fishermen in October on the northern North Sea coast of Schleswig-Holstein. Did they want to spawn there a second time?

Sagittate or Arrow Squid, *Todarodes sagittatus* (Lamarck, 1798)

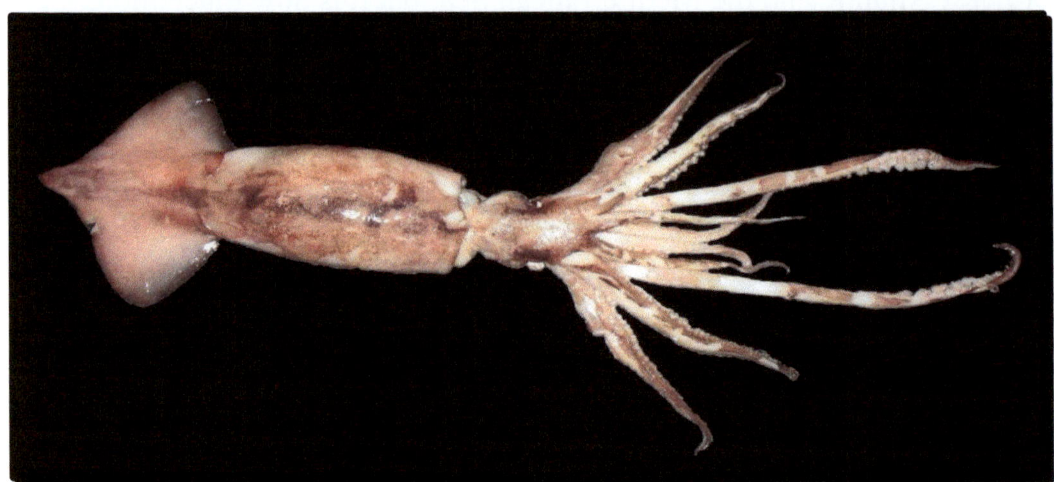

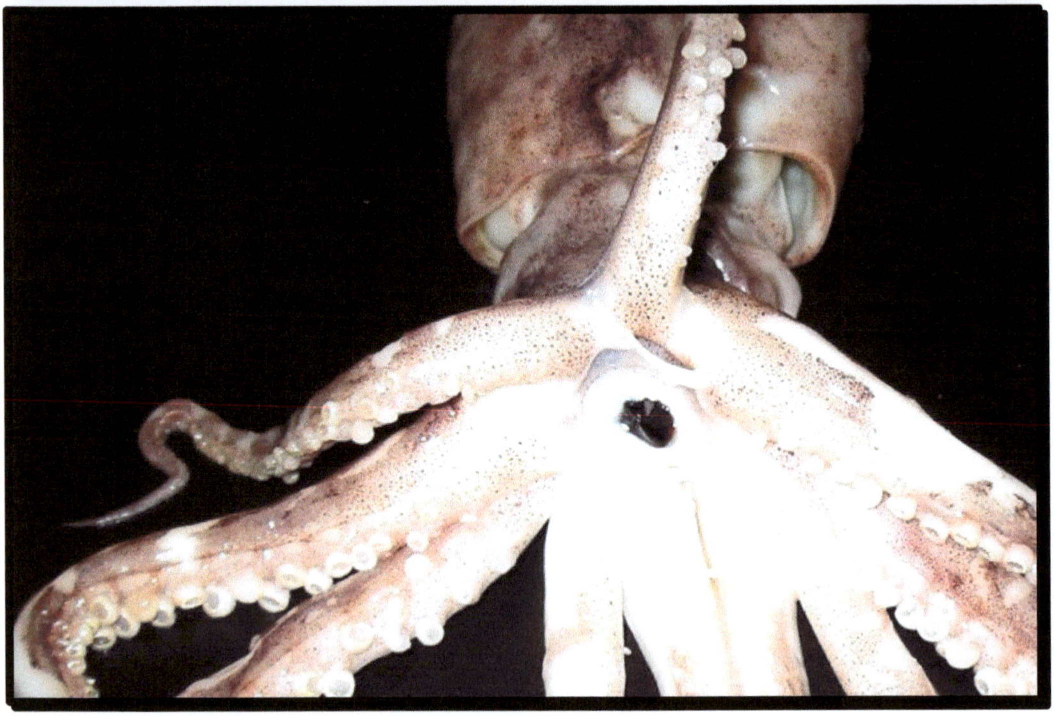

Beak of the Arrow squid. This is shaped like a parrot's beak and is suitable for chopping prey into bite-sized pieces. Big specimens could even cut off fingers or chunks of tissue with their powerful beaks…

The **Arrow Squid** is found in the Arctic, the North Atlantic and rarely in the North Sea or the western Baltic Sea. This species grows to a length of up to 150 centimetres and is thus already one of the larger species in this range. It is often found in the fish trade, where mostly smaller specimens from the South Atlantic are sold whole and then deep-frozen. The literature is somewhat contradictory to this species, because here it is listed under the synonyms ***Ommatostrephes sagittatus, Ommastrephes sagittatus, Loligo todarus,*** and even ***Illex spec.**.* In the edible fish trade it is also offered and marketed under these names. The exact species identification of squids is unfortunately anything but easy, but can be done by examining the shape and arrangement of the suction cups on the arms, the composition of the eyes and the shape of the fins at the end of the body. The shape of the jaws and the internal skeleton, the gladius, can also be included. Therefore, animals photographed alive are the most difficult to determine, as it is often not possible to examine these species-specific features in them. The specimen shown here on the right was caught by a shrimp-trawler in May 2016 in the southern North Sea near the East Frisian Islands. During the same period an even larger specimen was stranded on the northern beach of Borkum island, which can now be admired as a preserved specimen in the North Sea Aquarium Borkum. Because of its large distribution area, it is nearly impossible to say whether the appearance of such squids on our coasts is an evidence of global warming. However, the synchronous appearance of this rather rare species together with the **Forbes' Squid** could indicate, that a warming of the seawater has improved the reproductive conditions in our waters. This would allow the broods to develop more quickly and then, after hatching, move further northwards to feed and grow quickly.

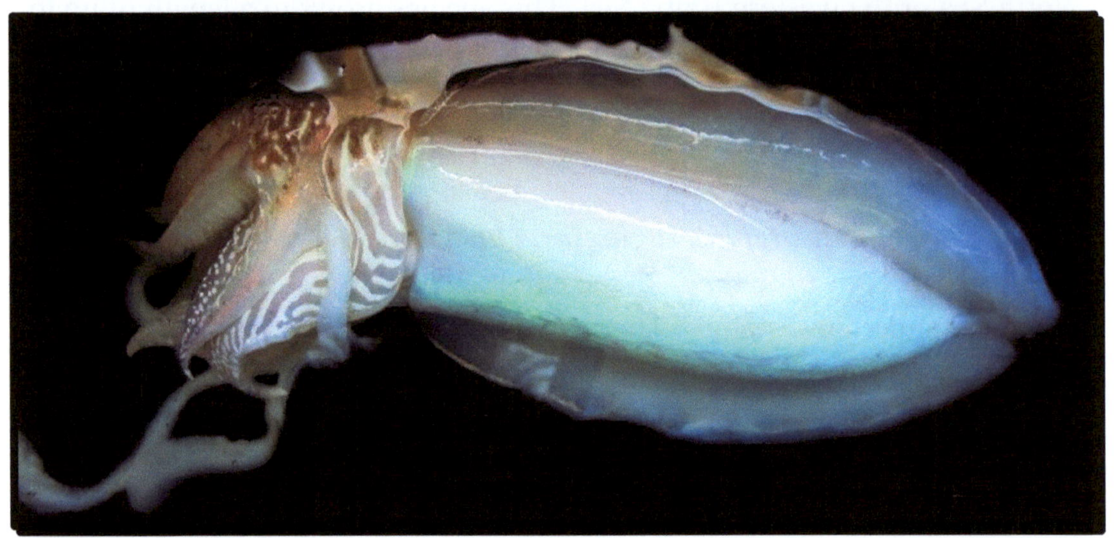

Common Cuttlefish, *Sepia officinalis* Linnaeus, 1758

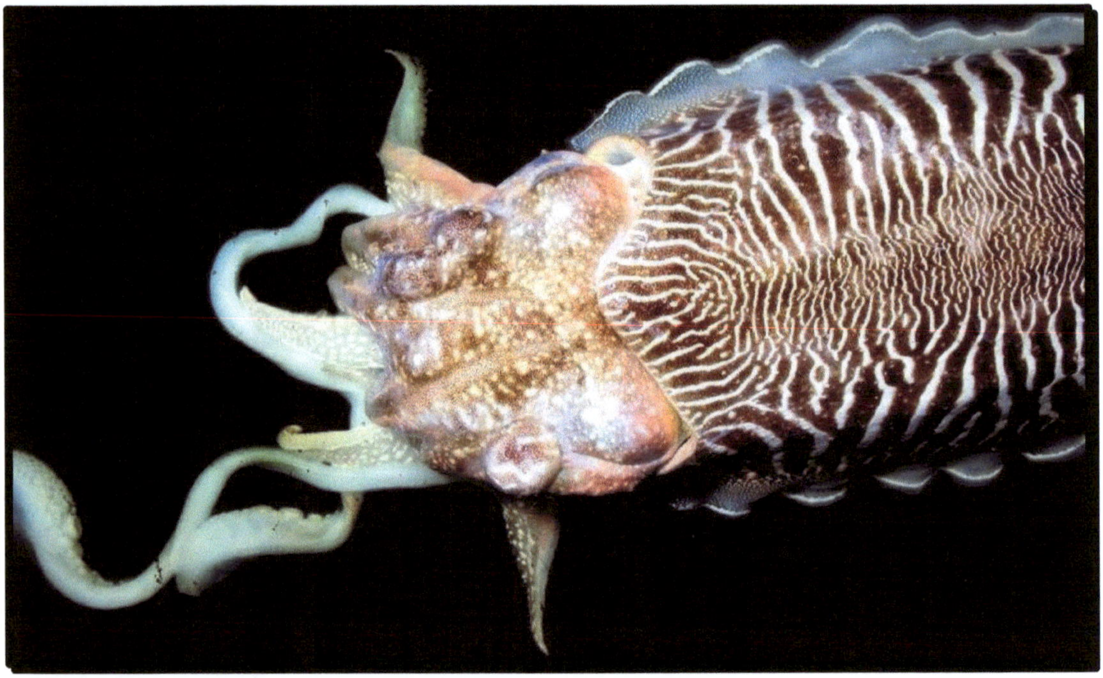

Common cuttlefish are able to glow with the help of symbiotic bacteria (bioluminescence). Note the iridescent fin seams!

Common Cuttlefishes can be found in the Mediterranean Sea, the Atlantic Ocean, the North Sea and the western Baltic Sea. They grow to a length of up to 50 centimetres and grow until they die a natural death after mating. The common cuttlefish seems to be seasonally common in the North Sea, as large quantities of washed-up internal skeletons of the squid, can be found locally on the beaches. These internal skeletons also bear witness to the fact, that cuttlefish in the North Sea reach quite respectable sizes. The classic spawning grounds for them are more in the southern North Sea, such as the mouth of the Oosterschelde and the English Channel, but cuttlefish egg clusters have also been found in the Dornumer Watt. As cuttlefish have recently been found in the Katinger Watt on the Eiderstedt peninsula, it is reasonable to assume that they are currently expanding their spawning grounds to the north of the North Sea. This can be seen as further evidence of the progressive warming up of the North Sea and should therefore give us cause for concern. If you want to keep cuttlefish, the best method is to hatch young specimens from egg clusters and then keep them in a species tank. This saves you the trouble of catching and transporting adult animals, which are very vulnerable here. They need relatively large aquaria, because in case of fright they can give off a recoil with the help of their breathing siphon, which transports them to the next aquarium pane. This may cause the cuttlefish's internal skeleton to shift and leads to fatal internal injuries. The cuttlefish can completely adapt to its environment by changing colour and body shape. It can activate small protuberances on the surface of its body, for example to look like a rough stone. It has also been observed by divers that cuttlefishes are very affectionate lovers who change colour during their mating depending on their mood. Cuttlefishes can bury themselves into the bottom substrate. There they lurk for small fish and crabs, which they prey on with their two long, pull-out tentacles. The prey is then chewed up and devoured with the parrot-like beak, which sits between the arms, into pieces. For their part, cuttlefishes serve countless predatory fish and also provide a protein-rich food source for humans.

These egg grapes were collected by a shrimp trawler in June 2016. On the left still with protective dark egg skin, on the right without this skin. They were attached to nylon strings...

When a shrimp-trawler collected these egg clutches in June 2016, it was unfortunately not possible to get living bent mysid shrimps (*Mysis spp*.) as food, and even frozen ones were simply nowhere to be found. The whole coast seemed to be swept clean, and so a successful breeding of the small cuttlefish was unfortunately not possible. This example shows how fragile the ecosystem of the southern North Sea has become. It should also be noted that the cuttlefish's parents had apparently laid their eggs on nylon lines. In plain language, this means, that the association with garbage in the North Sea's marine animals now begins in childhood. Thus, even the youngest organisms absorb the toxins attached to the garbage emitted by humans. As a result, these toxins enter the marine food chain earlier and earlier, which means, that the chain`s contaminated from the outset. People, who frequently consume such marine animals, themselves live in a very unhealthy way...

In the past, the two strains of ***Cnidaria***, i.e. cnidarians, and ***Ctenophora***, the ribbed jellyfish, were combined in the term ***Coelenterata***, but nowadays it is an invalid term. The literal translation of the word ***Coelenterata*** should actually be "intestinal holes". The tribe of ***Cnidaria*** is extremely diverse and has developed its greatest species richness in the tropical coral reefs. In fresh water there are very few representatives of this strain, but the freshwater polyps of the genus ***Hydra*** are especially hated by aquarium enthusiasts, who bring them into their aquariums with live food from ponds, where they then decimate the young fish broods. In the world ocean, the representatives of the cnidarians live in all water and depth zones, and with their planktonic larvae they are able to reproduce and spread over long distances. The free-floating medusas have sovereignty in all the world's oceans, as they dominate almost all layers of water and often mean the most common by-catch for researchers' fisheries in the deep sea. Some species in this group are very small and reach only a few centimetres in final size, while others can unite to form exorbitantly large superorganisms and reach lengths of up to 40 metres and more. Many cnidarians can use their nettle capsules to considerably nettle, stun, paralyse and kill their prey or enemies. However, it has only recently been proven, that some soft corals are even vegetarians who only feed on phytoplankton, because their nettle forces are not strong enough to hold animal plankton. The nettle process in all cnidarians can be imagined in such a way, that all cnidarians owe nettle capsules embedded in their skin. They content a small arrow, that is connected to a poison gland via a connecting tube. This capsule is closed with a lid, that opens on a stimulus. If, for example, the lid is stimulated by touching a feeding animal, it is opened in milliseconds and the nettle capsule ejects its poisonous arrow at high pressure. The toxicity varies considerably from species to species, so that only slightly poisonous cnidarians can even be eaten by humans. For example, various sea anemones in Scandinavian countries or on the Canary Islands are eaten by the locals. Or in Korea even jellyfish are prepared for human consumption. On the other hand, there are cnidarians, whose venom is also lethal to humans, because it usually leads to respiratory or muscle paralysis, which in water often causes the victim to drown. Especially feared here are the **Deadly Sea Wasp *Chironex fleckeri*,** as well as a cosmopolitan state jellyfish, namely the **Portuguese Man-O-War *Physalia physalis*.** Since even with stranded medusae the nettle capsules are often still active, you should not touch them under any circumstances! If you have been nettled after all, you should not simply rub off any nettle threads, that may still be sticking to the skin, but drip vinegar onto the nettled skin and scrape the nettle thread with a knife. It can also be helpful to powder the redness. In the case of severe nettle netting and allergies, you should consult a doctor as soon as possible. Fortunately, there are only a few jellyfish in the North Sea, which are painfully nettlers, but you should always be careful not to touch unknown cnidarians. After all, not all kinds of science are known here by a long way, and it is also conceivable that jellyfish from other sea areas have been carried into the North Sea by international shipping. Whereas in the past it was taught that coral reefs formed from lime could only exist in tropical regions where the water temperature did not drop below 20° Celsius all year round, this traditional doctrine recently had to be completely revised. This is because in the fjords of Norway, reefs with herma typical hard corals have been discovered, that are thousands of years old and which can be built without the help of sunlight at temperatures of just 4° Celsius. Their colours are in no way inferior to those of tropical corals! Also off the Icelandic coasts of the North Atlantic such reefs have already been proved, but unfortunately, they have been razed to the ground by the fishing industry by using heavy bottom trawls. That is why we are currently working hard to preserve the Norwegian reefs, which have just been discovered, for the posterity of our children by establishing protection zones. Since these reef builders are very slow to build reefs due to the prevailing low temperatures, the regenerative capacity of cold-water reefs is considerably lower than that of tropical or subtropical reefs. This is mainly due to the fact that metabolic processes of all kinds are much slower at low temperatures. Therefore, cold-water animals often become much older than warm-water organisms. Some of these organisms even become older than humans. For example, cylinder anemones of the genus *Cerianthus* have been kept in public aquaria for more than 80 years. And sea anemones of the species *Actinia equina* have also been successfully nurtured for several decades, with a single actinia spitting out more than 1,000 living juveniles, whose primordial tissue may still be circulating in public aquaria today. For this reason, such creatures are potentially immortal, because they can clone themselves using such propagation methods and their genetic material is preserved pure and unadulterated in their numerous descendants, who in addition still carry original tissue parts of their ancestors. Because cnidarians have always been drifted by the ocean currents around the entire planet, unfortunately very few conclusions can be drawn from the appearance and disappearance of individual species as regards of climate change. But the massive appearance of individual species in places where they were otherwise rarely found can often be taken as an indicator, that they are now finding more plankton than before. And the growth of marine plankton is of course accompanied by a rise in water temperature, because this promotes a more rapid growth of phytoplankton, which is then followed by the increasing crowds of zooplankton. The frequency and size of cnidarians found in relation to the season can therefore provide some indication of progressive climate change, even if these relationships are difficult to detect. It is particularly important to know, from which month onwards which species appear near the coast and how often. The class of **Jellyfishes (*Scyphozoa*)** is ideally suited for this type of studies. Because most of them are easy to find and to determine. And most of their spawning-products and larvae are very well known by the scientists and can easily be counted in aquatic tests.

Below are some typical jellyfish species that can be found from spring onwards in the southern North Sea:

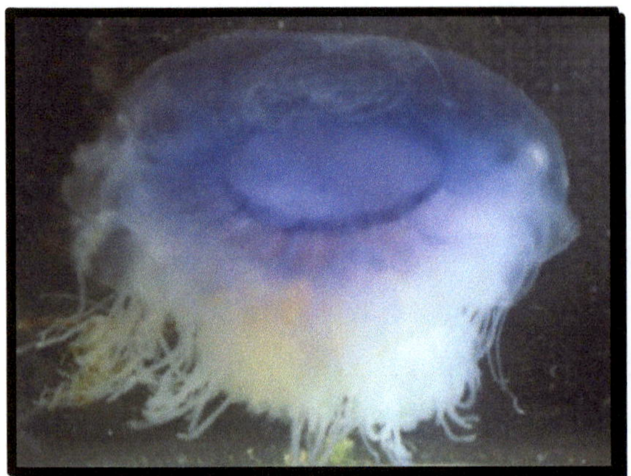

Blue fire jellyfish, *Cyanea lamarckii*, up to 20 centimetres radius.

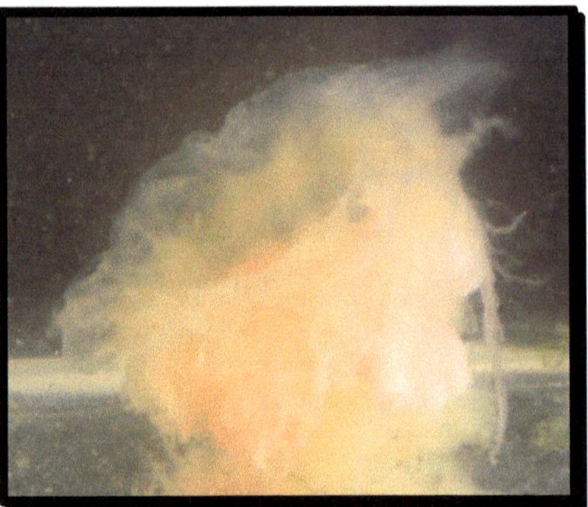

Lion's mane, *Cyanea capillata*, Up to 100 centimetres radius.

Moon jellyfish, *Aurelia aurita*, up to 40 centimetres radius. Locally very common.

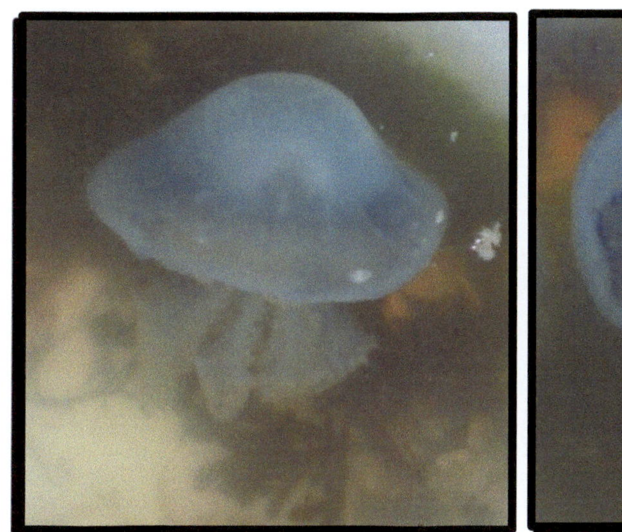 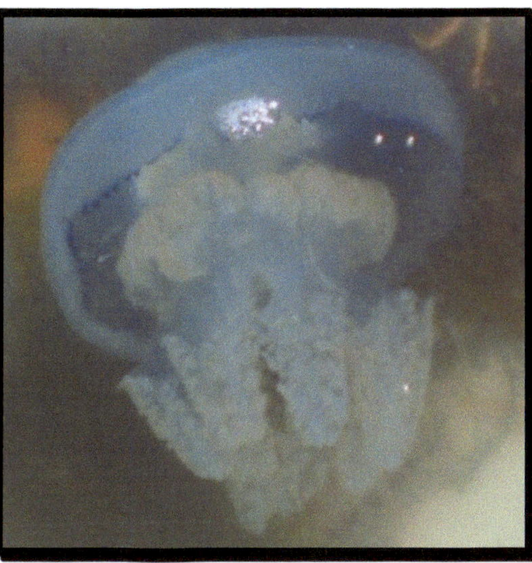

Football jellyfish, *Rhizostoma pulmo*, radius up to 30 centimetres. This species is more likely to appear in late summer in the German Bight. An earlier appearance could provide a warming of the sea...

Compass jelly, *Chrysaora hysoscella,* radius usually up to 20 centimetres. This jellyfish usually appears in the German Bight only in summer.

The small **Orange Anemone** originally came from the direction of the English Channel into the German Bight. It also occurs in estuaries with low salinity levels and can therefore be found in the Elbe estuary. It only reaches a diameter of about one centimetre and can easily be mistaken at first sight for juveniles of the **Plumose Anemone *Metridium senile***. But the plumose anemone has considerably more tentacles than the orange anemone. Such species are among the winners of the climate change. Since about the 2000s, this species has been advancing more and more. In the aquarium it is very durable and can be kept at room temperature without cooling. It remains to be seen whether the emergence of this species is faunal evidence of climate change, or whether it was simply carried into our waters by shipping. The fact is, however, that they can now be found all year round on the harbour pontoons in the southern North Sea. They are usually found in association with the shells of the also introduced **Pacific Giant Oyster *Magallana gigas***.

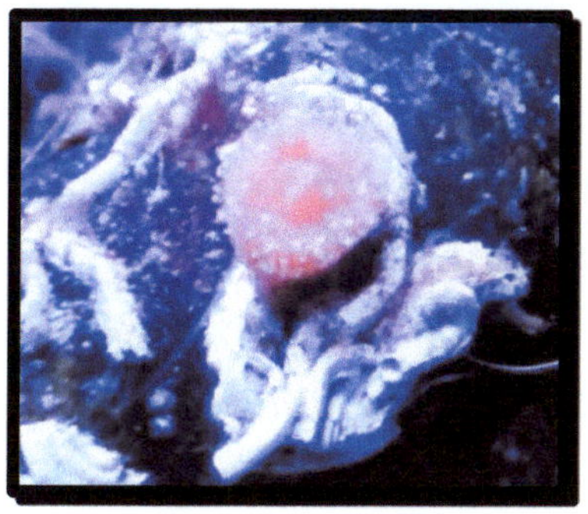

The **Cup Coral** can be found in the Mediterranean Sea, around the British Isles and in the North Sea, but especially on stony grounds. It is particularly well known from the British coasts, where it is also known as **Devonshire Cup Coral**. It only grows about one centimetre in size. The skeleton shown above was collected in 2016 by a German shrimp trawler off the Dutch coast; it was sitting between barnacles on an old fender. Since the cup coral usually does not find hard ground in the southern North Sea, it is most likely to be found on rubbish, shipwrecks or at offshore wind turbines. For these reasons, finding such a coral is a real rarity! Its skeleton is very similar to the tropical mushroom coral. Living specimens can be orange to greenish in colour. Since this real small hard coral belongs to the cold-water corals, its appearance cannot be taken as evidence of a climate change. For a long time this species has been known from the North Atlantic and the coasts of Norway. However, since it is often found in the English Channel in association with barnacle-species like ***Adna anglica***, it should always be checked when finding a cup coral whether these same barnacles are also found next to it. The recent appearance of these barnacles could well be interpreted as evidence of climate change, as they have so far only been known from British coasts. So I was able to find exactly this species of barnacle in January 2020 on a plastic box which had fallen into the nets of our fishermen off the island Juist. Unfortunately there seems to be very little known about how these animals can be kept in an aquarium.

Comb Jellyfish were systematically separated from the cnidarians as a separate tribe, because they do not have nettle capsules, but adhesive cells to catch prey. These in turn usually sit on two tentacles. In addition, they have fine lashes, which are arranged in symmetrical rows along their body axis. By moving these lashes they actively swim in the water, while the medusae from the cnidarian stem usually only drift or move by contractions of their whole body with the recoil principle.

Sea Gooseberry, *Pleurobrachia rhodopis* Chun, 1880

The **Sea Gooseberry** is very common in the shallow waters of the North- and Baltic Sea. It grows to about two centimetres long, but can extend its tentacles to at least ten centimetres. Sea Gooseberries are hermaphrodites, that hatch from the egg at about one millimetre in length. They are then already in sexually mature and reproduce immediately. Then their gonads first regress. When they are fully grown, their gonads develop for a second time and they can reproduce again. The eggs of adult animals are then considerably larger than those of the latest generation. If one studies their reproduction cycles over the year, one can draw conclusions about the warming of the sea. This is, because the higher the temperature, the more frequently it can successfully reproduce with even more generations, analogous to the plankton occurrence. Therefore, more attention should be paid to such animals in the future and they should be subjected to permanent monitoring.

Subclass *Errantia* – Non-Sessile Annelid Worms

The ***Errantia* subclass** includes representatives, who do not build living tubes or lime casings. These worms also have no body segments which are distinguished from other segments by other characteristics or special structures. Instead they have very well-developed stubby feet, which are called ***parapodia***. They also have many spines and bristles, which should protect them from predators. Some also have well developed eyes. Further species of this subclass are poisonous, and broken spines can cause nasty inflammations. To remove such fine bristles from the skin, it is best to use some adhesive tape, which is best for the spines to stick to. Some annelids also have strong jaws with which they can get a firm grip. They are not dangerous to humans, but you risk an unpleasant pinch in any case if you pick up such a worm. For this reason, rubber gloves should be worn to protect yourself when handling these worms. In this way you are protected from the fine bristles of these species. The size of the ***Errantia*** can vary greatly between a few millimetres in length and one metre, whereby such large specimens are rarely found near the shore. Many annelid worms, such as the representatives of the ***Nereididae***, are also regularly found on mussel beds or in between mussel clusters on groynes, where they hide from their predators during the day under the protection of the mussel webs. So you can often unintentionally introduce them into an aquarium with these mussel grapes. They are usually only seen at night when the aquarium lights are switched off and they go in search of food. The rule here is: The bigger such a worm becomes, the more predatory it is and can become a danger for other small invertebrates. Some worms are even so self-confident that they swim across the open water at night and obviously have no respect for a possible fish population. I could observe this behaviour in a worm that probably belonged to the genus ***Nereis***. From this I concluded that these worms must have very effective defence mechanisms against attacks from fish. Or it was in a mating mood and was looking for a sexual partner. The latter is a characteristic that is regularly fatal for many bristle worms. In February 2014, for example, I was able to observe a mass wedding of **Hediste diversicolor** on the beach of Neßmersiel, where countless worms were seen laying their eggs in broad daylight in shallow water. Some conclusions can be drawn from annelids and their populations about climatic changes, if one carefully studies their populations and their reproduction habits over several years, better still decades. For example, there are species whose populations die completely in cold winter years during extreme cold spells, thus disappearing from the southern North Sea. This means that after such a cold shock, these species must first leave the English Channel and head north again to reclaim habitats here. They do this by entrusting their **Trochophore larvae** in the southern areas into the Gulf Stream, which then carries them northwards into the German Bight. Here they then develop into adult worms and, under the otherwise actually "normal" environmental conditions in the German Bight, would die off completely as a population in winter and thus serve as food for other organisms. If, on the other hand, the winter remains "warm", they no longer die and begin to reproduce. What effects this has on the ecosystem of the southern North Sea can only be assumed. Another effect of the warming up of the North Sea water is, that the annelids change their life cycles and start reproducing earlier and earlier in the year. This then has many effects on other animals such as seabirds, migratory birds and fish, which feed on these worms. And are dependent on their presence at certain times of the year in certain locations. We can be very curious to see which worms will occupy which ecological niches in the future, and when they will do so. This may then have serious consequences for bird migration and fisheries, which will then, out of the blue, run into problems that were previously unknown. Such as crowds of not usable by catches and a lack of shrimps and fish, which have been once very common species. And a scenery with a lack of sea-birds at the one hand side and a coast decorated with thousands of starved migratory-bird species on the other hand side…

The **Catworm** reaches a length of up to 20 centimetres. This species is found not only in the North Sea but also in the Black Sea and the Mediterranean. This is why this worm is one of the thermophilic species that used to have its northernmost distribution limit in the southern North Sea. However, it is now also found in the Skagerrak, Kattegat and the Baltic Sea. In cold winters, populations of this species can die completely, but they can compensate for their losses the following spring with higher reproduction rates. Catworms are strong predators, which feed on juvenile mussels, small crabs and other worms. In 2013, despite the long winter, I was able to obtain a specimen of this species from a shrimp trawler. The cold spell lasted until April 2013 and caused the plankton and life cycles of the North Sea animals to shift backwards by at least one month. Obviously, this cold winter did not have any effect on the population of these worms in the 2013 season, and in autumn 2019 I even found some juveniles, what proves, that the catworm is reproducing in our coastal waters. So this species has obviously already established itself successfully in the German Bight due to global warming.

King Ragworm, *Alitta virens* (Sars, 1835)

The **King Ragworm** is a very hidden annelid that can grow up to 80
centimetres long. During the day, these worms can hardly be seen, as they
live in a firmly cemented tube. Only at night they are brave enough to come
out of their tubes and eat algae, detritus and small invertebrates. With their
jaws they can get a strong grip, and if you touch a worm the wrong way, it
can pinch. With their paddle feet they are elegant swimmers, which can
quickly snake through the water. In April and May these worms reproduce
by swimming to the surface. This means a rich laid table for various seabirds
and fish. In April you can find strange transparent greenish lumps (see
picture on the right) in the waddens. These are the egg packs of these worms,
which until they hatch are in a transparent jelly mass similar to a grape. After
the winter 2019/2020 had obviously "failed" in the southern North Sea, the
egg clusters of this sea annelid could be found in the tidal flats as early as
March 2020. If this trend continues, it is even conceivable that such worm
species could reproduce several times in just one warm season, so that such
green egg grapes could be found in the waddens in autumn. We can be
curious about the phenomena we will observe here in the near future...

The **Sea Mouse** is a relatively large annelid that can grow up to 15 centimetres long. It is usually found from a depth of 10 metres well below the tide line and only very rarely after stormy days in the rinse zone. The sea mouse hunts small animals like molluscs and worms in the mud, but also feeds on carrion and detritus. After observations in the aquarium, it can be said, that this species prefers a cold environment and dies quickly, especially if it warms up too quickly. Therefore, this species could also be used for a targeted monitoring of climate change by checking how deep you have to fish to find living healthy specimens. Furthermore there are some places, were dead specimens are washed up on the beaches regularly. So here exists the possibility, to count these specimens in relation to the developed water-temperatures.

The *Crustacea* **subphylum** includes more than 40,000 known species and more new species are described daily. The crustacean class includes several species of lower and primitive crustaceans, such as rudder crabs, phyllopods or parasitic rhizocephalans; also barnacles, goose-barnacles and several others. However, families such as shrimps, crabs, lobsters and crayfish are more common. Crustaceans have been referred to as the "insects of the sea", and in fact, from a purely ecological point of view, this view is not all that wrong. This is because crustaceans serve as food for many other animals in marine ecosystems, while they themselves often occupy the ecological niche of the health police. Even the large baleen whales feed on tiny crustaceans, such as Arctic krill, and could not survive without this important source of protein. Many crustaceans are also an important source of protein for humans, but some species have also become simply luxury items. Entire fisheries and secondary industries depend on certain crustaceans and the quantities landed, such as shrimp-, lobster- and crayfish-fishing, but also other fisheries for swimming crabs, big prawns and edible crabs. And, of course, the fisheries for herring, plaice or cod also depend on tiny crustaceans, as these are often the food source for these commercially important fish species. In the southern North Sea, i.e. on the East Frisian Islands and on the beaches of the mainland, some disturbing and sometimes contradictory phenomena have been observed in crustacean stocks for several years. First of all, as if from nowhere, species are suddenly present here that were previously unknown from this area. They were mostly introduced to Europe by shipping. These include species such as the Pacific shore crab or the oriental glass shrimp from Korea. These species have long since earned a place for themselves in the Wadden Sea ecosystem and have been able to adapt to the environmental conditions prevailing here, also due to the new warmer environment. Then there are other species whose distribution area extends from more southern areas in the English Channel further and further northwards. These include glass shrimps, dwarf hermit crabs, velvet crabs, several swimming crabs, the great Atlantic spider crab and also shore crabs of the Mediterranean Sea. Another problem is that some immigrant species can also mate with native species. Hybrids are then formed, which cannot be clearly assigned to any species. This has probably already happened with the common shore crab! Unfortunately, it is currently unclear whether these new hybrids will be able to reproduce fruitfully. By the way, this would also be conceivable from the introduced glass shrimp of the genus *Palaemon*. And then there are the autochthonous species of the North Sea, some of which are now declining and threatened with extinction in the southern North Sea. Such as the great spider crab, for example, which belongs to the Arctic faunal group and will therefore probably disappear from this part of the sea in the foreseeable future. Such species then migrate northwards. But also the common true shrimps, on which an entire fishing branch rests, have already started to migrate with the flow. Which sometimes leads the shrimp fishermen following these animals with their boats further and further northwards unto the coast of Schleswig-Holstein. And then there are even shrimp species, which actually can be used as a "climate oracle". Especially the **Aesop Shrimp *Pandalus montagui***, which is a distinct cold-water species. It is occasionally found at weekly markets among cooked true shrimps. When this species is caught, the catch either comes from depths of more than 20 metres, or it announces the approaching autumn or winter. If these shrimps are no longer caught, summer is approaching. And if one warm day they are no longer present in the southern North Sea, this means, that the North Sea must have heated up to more than 14° Celsius all year round, and that all efforts to protect the climate have finally failed... In the following we will take a closer look at various crustaceans and their relevance for statements about global warming in the southern North Sea. Some results from the frequency, others from the mere presence or size of the specimens found. It is not possible to make sweepingly simple statements here. Because everything is part of a larger context. But if we combine some of that observations, we will discover some developments during the last decades, which only can be called still frightening…

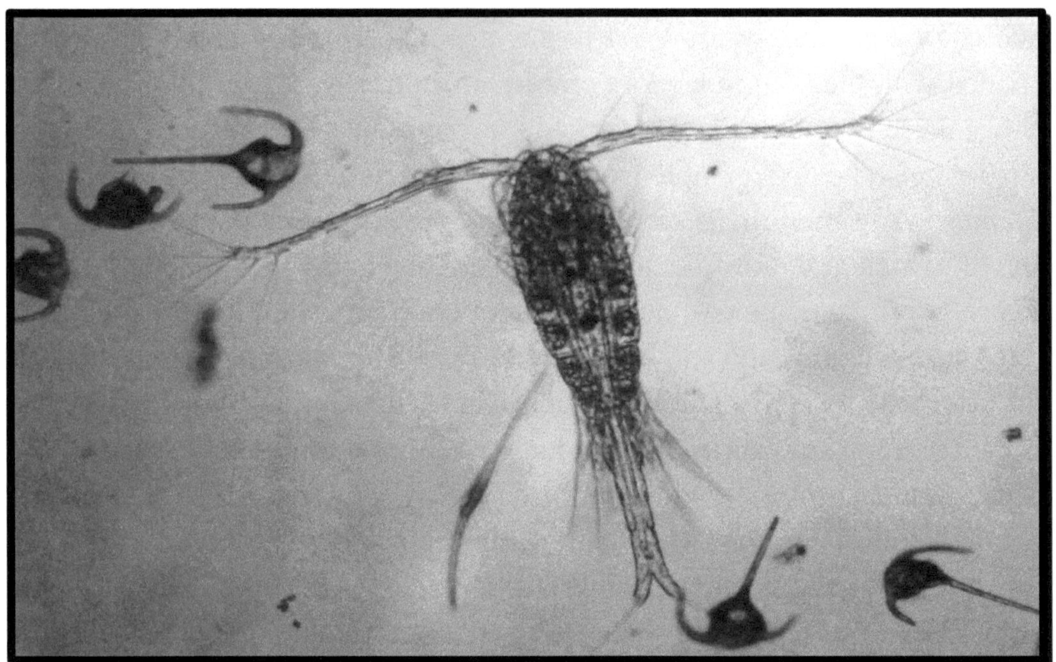

Rudder crab, *Calanus finmarchicus*, a common species of the North Atlantic. The copepods of this genus can reach a length of up to 10 millimetres.

This copepod is a very important fish food in the plankton column of the North Atlantic. The water temperature influences its vertical migration cycles in the water column. This means that the development and welfare of our fish stocks depend on these small crustaceans. When disturbances occur, entire fish populations can migrate or collapse in the short term... It is also often the case, that the larvae of fish or other commercially important crustaceans specialise in specific prey in the plankton. This means, that they cannot simply orient themselves to other food as a result of a warming of the environment. In the worst-case scenario, they will starve to death or only decimate each other, as it is the case with the larvae of the **European Lobster *Homarus gammarus*.**

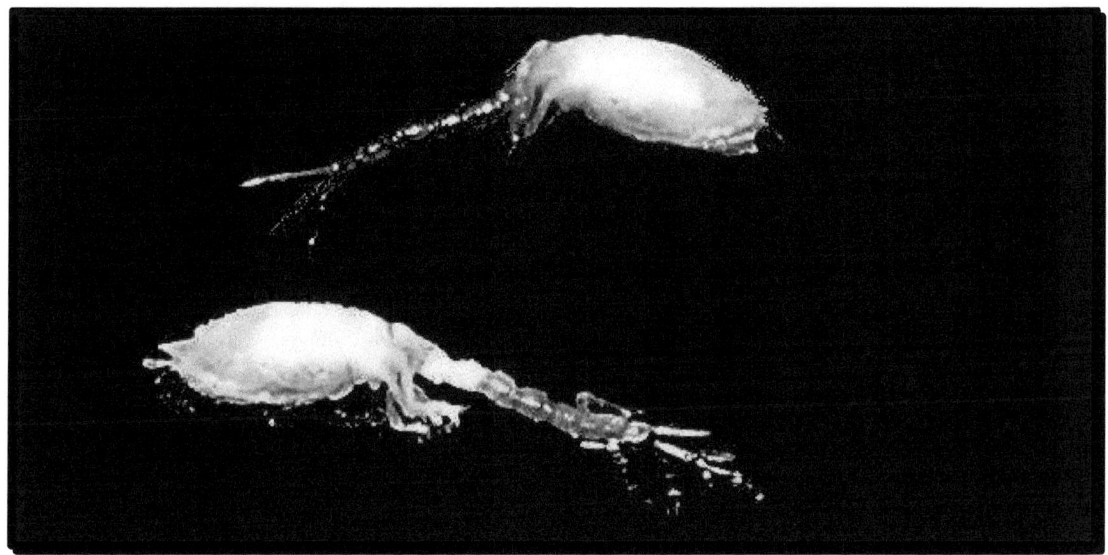

Rathke`s Hooded Shrimp, *Diastylis rathkei* (Krøyer, 1841)

Hooded Shrimps are already somewhat larger than most copepods and are counted among the so-called macrobenthos. This means that they are small crustaceans, which reach a size of about one centimetre. Hooded shrimps live on the sea floor, where they feed on detritus and leave feeding channels in the mud. At night they rise up in the water column towards the surface and become important fish feeders. In the Kattegat, it was already noted in the early 2000s, that their stocks had collapsed by 70%. Now, years later, in 2020, herring quotas have been reduced by 65%... Is there a connection? Unfortunately, it is not clear at this stage whether these animals have already become victims of climate change or whether they are just suffering from contaminations of the seabed. It maybe, that both possibilities are simply true.

This infra-class of crustaceans includes both: Barnacles and the somewhat less well-known goose barnacles. That they are crustaceans can no longer be seen in adults, but can be seen in larvae before their metamorphosis. It is difficult to evaluate these organisms as climate indicators, because as early as the 1940s, the shipping of the Second World War caused barnacles to be transported from other parts of the world to the North Sea. The Australian barnacle ***Austrominius modestus***, which can still be found in the German Bight today, is the best-known example of this. It came from Australia as a result of troop transports to Europe and has successfully established itself here permanently. The goose barnacles, on the other hand, are cosmopolitans, that can be found in all oceans. They appear sporadically in large clusters in certain places and are spread worldwide by the ocean currents. Therefore, their appearance in other parts of the sea cannot be taken as evidence of a climate change. But their increased occurrence in the German Bight could be evidence of an unusually good reproduction of their planktonic nutrients as a result of the warming up of the North Sea water.

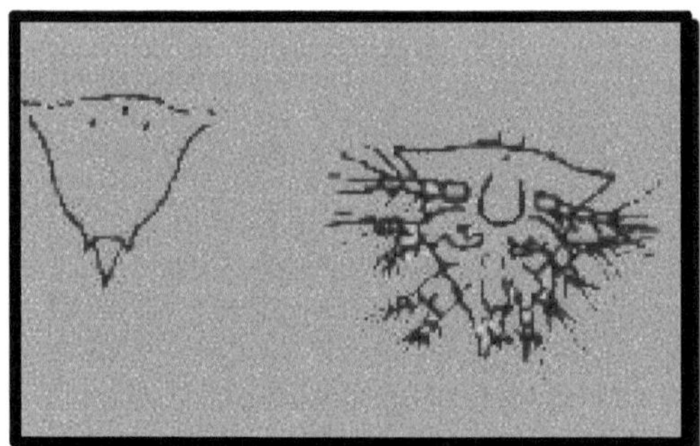

Nauplius **larva of barnacles and** *Cirripedia* **larva of goose barnacles. In these early stages of their life they owe eyes and legs. And furthermore they are able to swim and to move actively to new substrates, where they settle down and become sessile animals by a final metamorphosis.**

Goose Barnacle, *Lepas anatifera*, 40mm. Australian Barnacle, 12mm.

***Perforatus perforatus* and *Adna anglica*, 5mm.**

These two barnacles are a novelty in the southern North Sea, which I could find on plastic for the first time in January 2020. Until then these species were only known from British coasts. To the species ***Adna anglica*** it should be noted, that it is usually associated with the **Cup Coral *Caryophyllia smithii***. In this context it is useful, to take a closer look at accumulations of barnacles, because with a bit of luck it is possible to find some rare cup corals. Because of their cone shape, these two species of barnacles are sometimes also called "volcanic" barnacles. Another characteristic feature is the delicate purple hue of these species.

These crustaceans are known from all the oceans of the world and also from brackish and freshwater habitats. Some species have been spread worldwide by man, while others simply follow the warm Gulf Stream northwards. Therefore some of these species are very good indicators of recent climate changes. A common feature of all decapods is, that they always have ten pairy legs, if you count their pincers, too.

Chamaeleon Prawn, *Hippolyte varians* (Leach, 1814)

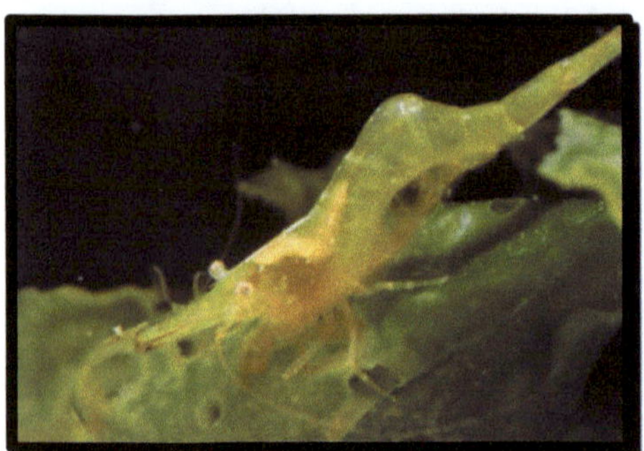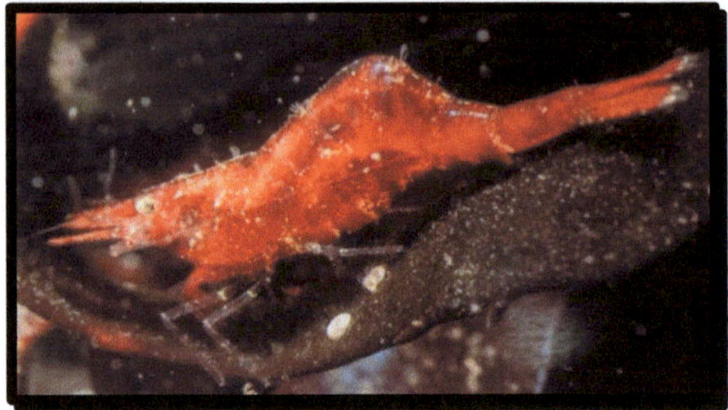

This specimen of the **Chamaeleon Prawn** was found 2016 in the harbour of the island Borkum. Seagrass-shrimps are actually more familiar from the Mediterranean or the Irish Sea. They are otherwise found regularly unto the Dutch coast. Their distribution area is comparable to that of the **Dwarf Hermit *Diogenes pugilator***. All they need is suitable seaweed, hard substrates or seaweed stocks. And some kind of warmth! The two photos above show one and the same specimen. They adapt to their environment and can change colour as desired or even become completely transparent. Because of their small size they are hardly noticeable. Therefore it is also conceivable that they have been in our waters more often in recent years, but nobody has noticed them here yet. Moreover, they are so filigree that they can very probably almost always slip through the meshes of the fishing-nets. But they can also be understood as a very small, discreet warning from Mother Nature to us humans to change our climate-damaging behaviour.

Processa-Shrimp, *Processa canaliculata* Leach, 1815

The **Processa Shrimp** is a small and inconspicuous species, which could be found sporadically from 2014 onwards. Although it is widely distributed from the English Channel to Norway, it has hardly been found in the German Bight so far. From autumn 2018 and autumn 2019 onwards, more and more specimens were found in the by-catch of our shrimp fishermen. Whereby in October 2019 there were about ten times as many as in the previous year. With lengths of about 30 millimetres, however, they were well behind the sizes they can reach in warmer areas. This is also the reason why they do not end up in the fishermen's cooking pots together with the common true shrimps. Because of their small size, they are simply sieved out beforehand. Otherwise, in my opinion, they could have been found as by-catch in previous years. It was only when I took the trouble to examine sifted shrimp from a shrimp trawler, that I noticed this usually slightly reddish shrimp. The mere appearance of this species does not directly prove the climate change in the southern North Sea, but it does prove that something must have changed in favour of this species. Otherwise the juveniles of this strange shrimp would hardly be able to develop in our latitudes in any appreciable numbers.

The **Oriental Glass Shrimp** *Palaemon macrodactylus* originally came to our waters from Korea and Japan. Her body is transparent to beige and brownish in colour, with a characteristic dorsal stripe in the middle. The species was most likely introduced into our waters in kind of larval forms with the ballast water of ships from overseas via the cargo port of Rotterdam. However, this glass shrimp seems to prefer brackish water with slightly lower salinity, as it is found mainly in estuaries and ports. It is remarkable, that it has obviously adapted to our climate and is therefore also proceeding to reproduce in our waters. In some places it can seasonally push back native glass shrimps and even dominate a habitat as a species. The fact, that this species is now reproducing regularly in our harbours, can be seen as evidence of both: The species' ability to adapt to new water conditions and to the warming up of the North Sea as a whole. Otherwise, species such as these would die off during the cold winter-period.

Autochthonous glass shrimps of the genus *Palaemon*

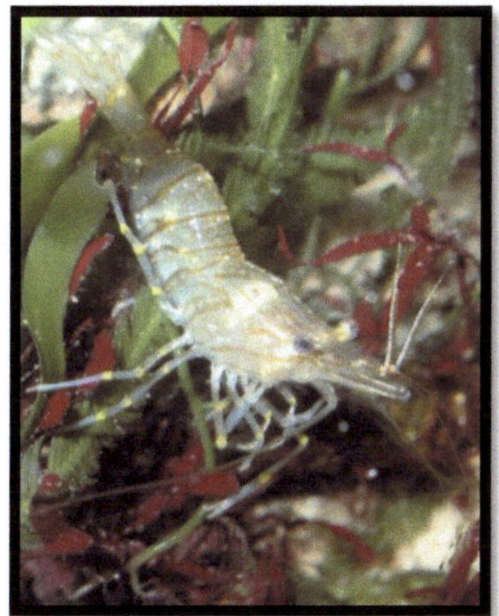

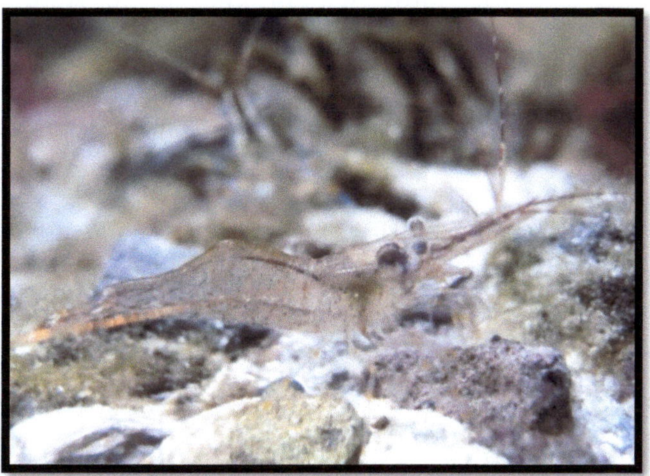

Glass Prawn, *P. elegans* 60mm.

Baltic Prawn, *Palaemon adspersus,* 50mm.

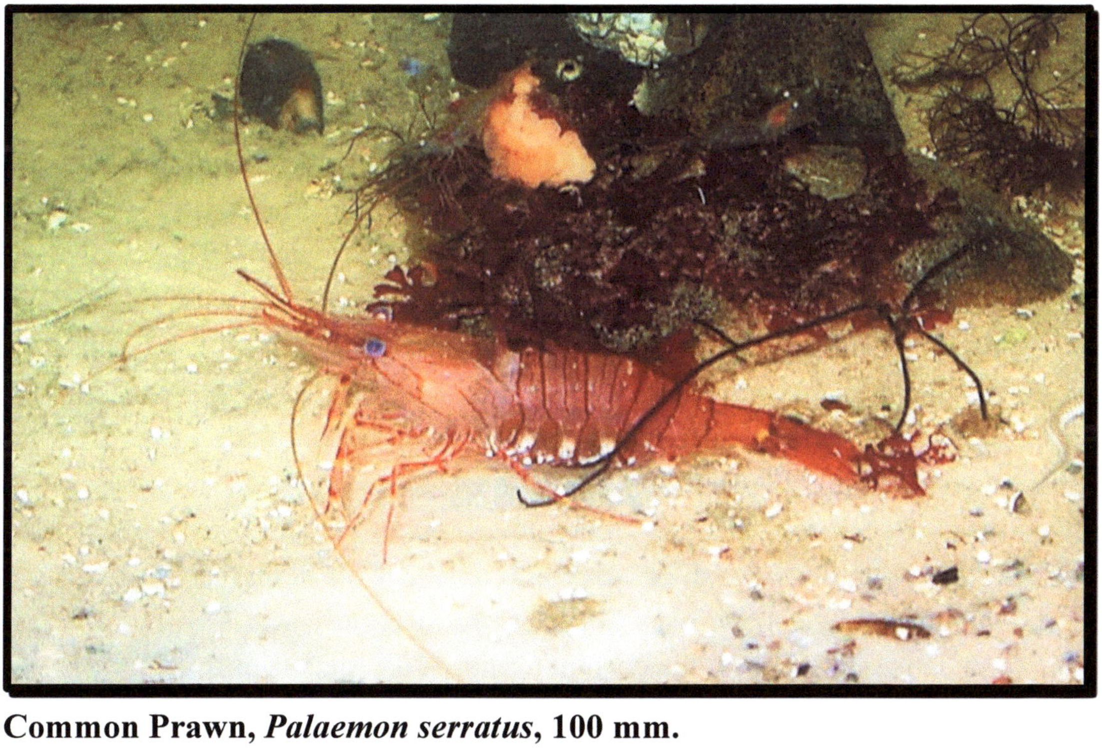

Common Prawn, *Palaemon serratus*, 100 mm.

The **Glass Prawn *Palaemon elegans*** is a very common shrimp found in the North Sea, the Baltic Sea and also in the Mediterranean. It has probably been spread halfway around the globe by ships in whose ballast water its larvae travel. These animals grow to about 6 centimetres in length and prefer to sit on sheet piles, on stones, in sluices and on piers. They are small agile omnivores. The females can be recognised from a size of about 4 centimetres by the larger abdominal shields of the abdomen, the males by the smaller segments. The females deposit free-swimming Zoëa larvae, which after a planktonic phase of about 3 weeks transform into finished small shrimps. The pincer legs of this species are curled blue-yellow, and the movable *dactylus*, i.e. the pincer finger, is at the bottom of the pincer. And not on top, as in most other crustaceans. When the water is particularly warm, they get little yellow dots that cover the entire body. In the meantime, new species of the genus *Palaemon*, such as the **Oriental Glass Shrimp *Palaemon macrodactylus,*** have also been introduced to Europe by ships. Since these species all look similar, the only way to determine the exact species is to count the number of rostral spines, measure body proportions or carry out a molecular genetic analysis. The **"Baltic shrimp" *Palaemon adspersus*** is widely distributed, because it is regularly found in the Black and the Mediterranean Sea, the southern North Sea, around the British Isles, in the Baltic Sea and in the north as far as Norway. In addition, it is now also known from the coasts of India, and probably thanks to international shipping, it has now opened up various other parts of the world's oceans for itself as a habitat. It has two to six teeth on the upper part of the rostrum, while the lower part has only two to four teeth. Like some other species of this genus, it has the typical blue-yellow curled legs and pincer legs. Its body colouring can be very variable. Depending on location and season, there may be single-coloured brownish or beige animals, as well as bright yellow or even slightly orange specimens. Shrimps produce these stains with the pigment cells in their skin, the chromatophores. Baltic shrimps also tolerate low salt levels and higher temperatures, which is why it is easy to keep them.

The **Common Prawn, *Palaemon serratus*** is very common and can be found from England and Denmark in the north unto the Mauritanian coast in the south, throughout the Mediterranean Sea and near the Azores. It can reach final sizes of 11 centimetres and more. It is very difficult to distinguish their young specimens from the **Glass Prawn *Palaemon elegans***, which is why the rostral spines should be counted in smaller animals. The **Common Prawn** prefers rocky substrates, which is why it cannot be found in the waddens. However, it is occasionally caught by shrimp-trawlers. This species is mostly landed as a single animal in German waters, so that commercial use is not worthwhile. In the Mediterranean, Portugal and Brittany, this shrimp is also used as a food shrimp, as larger quantities can be caught here regularly. This shrimp is said to belong to the nocturnal species and has a similar way of life to the glass prawn. It can be kept just as well in the aquarium, but requires slightly colder water, which should not be warmer than 18° Celsius. The 100-millimetre-long female specimen shown here was caught by a shrimp-trawler and had bright red crossbands and blue eyes. Possibly this is an adaptation to the red rocks of Heligoland, because the specimen had been caught near this island. Finally, with regard to our native glass prawns, it should be noted that glass prawn and Baltic prawn do benefit from a warming up of the North Sea. This is, because in a warmer environment their larvae can develop faster than before, if the phyto-plankton reproduces faster, too. It may happen, that they produce several new generations of juveniles per year. It can already be seen, that their presence on the sheet pile walls of small ports is associated with rising water temperatures. However, there is also an upper limit of an estimated 20° Celsius from which they retreat into cooler deeper water layers. For the common prawn it can be seen that adult specimens do not tolerate temperatures above 18° Celsius for long. On the other hand, their larvae should benefit from a higher plankton supply. Thus, we can expect that this species will be preserved in somewhat deeper and oxygen-rich areas in the future. Especially near offshore wind turbines such as at Borkum island.

The **Aesop Prawn** is a proof for that a mere reddish colour does not necessarily indicate a subtropical species. In fact, the opposite is true. After all, this is a shrimp of the Arctic faunal circle, and beyond the 14° Celsius mark this species dies in days. If it becomes permanently too warm, it will therefore simply "disappear" from the southern North Sea. This species is still regularly caught by our fishermen as by-catch during the cold season. From this you can then either draw conclusions about the depth at which it was fished or whether winter is approaching. If they are no longer caught in shallow water, then summer is approaching. And if one warm day this species is no longer found at all in the southern North Sea, this proves, that all the climate protection efforts of mankind must be considered a failure. This species in particular should therefore be subjected to constant monitorings and surveillance of its habitats. It can be taken as an indicator species for the increasing temperatures in the North Sea.

The **Shore Crab *Carcinus maenas*** is the most common crab of the German Wadden Sea and the Baltic Sea. Males can reach a carapace width of up to 8 centimetres, the females remain somewhat smaller. During the ebb tide they can be found in tidal pools or under stones, where sometimes up to 100 specimens crouch and wait for the next flood. They tolerate extremes such as salt density fluctuations, heat and cold without problems. If they did not find cover, they simply hide in the sand. Shore crabs are predatory and eat whatever they can find or overpower. Therefore this way of life can be called opportune. On the one hand, the shore crab takes the part of a health police in the Wadden Sea, but on the other hand she itself means an important food source for seabirds and certain fish species. In winter, the shore crabs migrate to deeper waters off the coast, as the waddens become too cold for them. In spring the larger shore crabs usually reappear in the waddens in larger numbers. In the meantime, the shore crab has been spread all over the world by ships, so that it can now also be found in the harbour of San Francisco, for example, where it is growing bigger than in the North Sea and reaches widths of up to 10 centimetres. The shore crab is thus a beneficiary of globalisation and climate change and is therefore likely to survive in the Wadden Sea in the future despite the introduction of other crab species. At present, however, it is unclear whether the shore crab will continue to exist as a purebred species, or whether there will only be hybrids of **Shore Crab *Carcinus maenas*** and **Mediterranean Green Crab *Carcinus aestuarii*** in the future, as the two species mate with each other.

This species belongs to the same family and genus as our native shore crab. It was introduced into our waters from the South Atlantic and the Mediterranean Sea. The **Mediterranean Shore Crab** has no rostral spines at all between its eyes. In addition, the males' mating organs are straight and parallel, whereas the native shore crab owes curved mating organs. However, it mixes with the shore crab, so that in the meantime one does not know exactly, which species one is looking at. Especially because some specimens show characteristics of both species. Whether these hybrids can continue to reproduce is very probable, but not yet clear. The specimen shown here was collected in Wilhelmshaven in June 2008. Since the Mediterranean shore crab was not yet known from the North Sea in the 1970s and 1980s, it can now be considered to be an evidence of climate change among some further crab-species.

This species first entered the German Bight during the hot summer of 2018, where a shrimp trawler had caught it off the island Juist. Until then, finds of this crab from the German coasts were completely unknown. As this species was only known from the southeast Atlantic, the Mediterranean Sea and the English Channel, the northernmost extension of its occurrence. The summer of 2018 was extremely dry and hot, and the surface temperature in the southern North Sea was about 22° Celsius. While the river Rhine had warmed up to 28° Celsius in places at low water levels, and water temperatures of up to 25° Celsius had even occurred in the southern Baltic Sea. At first sight, this crab resembles a swimming crab, but it lacks the typical flat webbed feet. Therefore it belongs to the same family as the shore crab and the Mediterranean green crab. It is an absolute novelty in the German Bight and thus also a good and objective proof of the consequences of climate change in our coastal waters. The eon of ***Xaiva*** has begun...

Linaresi`s Long Legged Spider Crab reaches a length of about 20
millimetres without legs. This species was only recognized as a separate
species in 1964. In recent years it has been caught frequently by our
fishermen. Mostly on plastic waste or the remains of nets or nylon threads. It
is otherwise well known from the Mediterranean and North African waters...
It is therefore obvious that this species has successfully conquered new
habitats in the north due to increased temperatures, and has already
established itself here. It can be kept at room temperature in an unheated
aquarium very well.

This species from the large superfamily of spider crabs is mainly known
from the English Channel and the Canary Islands, where it is also fished
commercially in large numbers. Since the beginning of the 2000s, it has also
been known from occasional catches from Heligoland... (It has also often
been confused with the **Mediterranean Spider Crab *Maja squinado***). With
the rising temperature in the southern North Sea, it would not be surprising,
if they were to be found more frequently in the future. Associated with the
new large offshore wind energy plants and sitting on the large mountains of
waste on the sea floor. If small juveniles of this species are then also be
found, it means, that the species has established itself and is also starting to
reproduce in the North Sea.

The **Marbled Swimming Crab**, which is only about 2 to 3 centimetres in size, was also previously known from the English Channel and British waters. Now it appears occasionally in the southern North Sea. It can be kept in a sea aquarium with sufficient water circulation at room temperature without any problems. Even without having to operate a cooling system! A special characteristic of this species is the subtle marbled pattern on its shell, which is an excellent camouflage especially on gravel and scree. Furthermore, the shape of their pincers is typical for this species. Such swimming crabs can sometimes also be found as cooked by-catch between **Common True Shrimps (*Crangon crangon*)**. However, the patterns on the body then no longer look brownish but reddish. This may be a reason, why that swimming crab has not been remarked earlier in the German Bight.

The **Arch-Fronted Swimming Crab** has been a frequent by-catch of shrimp fisheries for several years. It often appears in the southern North Sea at the beginning of the season in March or April. This small swimming crab is usually only two or three centimetres large and can therefore be easily overlooked. Especially typical for this species are the missing rostral spines between the eyes, which is why we can also speak of a smooth rostrum here. Its body colouring is also white-brownish marbled, which is why it can be confused with the **Marbled Swimming Crab *Liocarcinus marmoreus*** at first glance. However, the latter has several thorns on the rostrum between the eyes. The arch-fronted swimming crab is now very widespread and was formerly known mainly from African coasts and also from the Mediterranean Sea. Due to the shipping traffic it seems to have become a cosmopolitan species in between.

The **Velvet Crab** can reach quite considerable sizes and develop a leg span of well over 10 centimetres. This species used to migrate temporarily from the English Channel into our waters as an adult crab during the summer. Now it seems to stay here and also hibernate. In the spring of 2020, for the first time, the shrimp trawlers of Norddeich (Lower Saxony) brought back not only adult velvet crabs, but also juveniles from 2 to about 4 centimetres in size. The velvet crab has fine bristles all over its body, which actually feel like velvet. Another typical feature of this species are the red eyes. (And because of this attitude, this crab is sometimes named as "Devil Crab")! The velvet crab is commercially bred in Spain as an edible crab. It also lives at the offshore wind turbines near Borkum island and migrates from there to the north, where it then settles comfortably on the mountains of plastic waste and net remains to hunt for the unfortunate small animals, which regularly get tangled up in the rubbish...

Already in 2003, I was able to pick up this crab during an Easter holiday on Baltrum island after a storm surge as a small stranded swarm on the beach. In 2014 this crab appeared in April in large flocks. Later in the summer on Norderney it was even possible to catch it in the shore area. In between it comes from Northwest Africa to the southern North Sea to reproduce here. In warm years they are very common locally. But if it is too cold, it stays away completely. Therefore, such species are good indicators of the local climate in the North Sea and deliver information, about how the summer will develop. This species must be regarded as a reliable climate indicator because of its original distribution and origin in Northwest African waters.

Strigose Squat Lobster, *Galathea strigosa* (Linnaeus, 1761)

The **Strigose Squat Lobster** is a very abundant species, because it is found in the Mediterranean Sea as well as in the North Atlantic. Some specimens have even reached the Red Sea through the Channel of Suez. In Europe, specimens of this species have been recorded from the coasts of Belgium, Norway, Sweden, Ireland and the British Islands. From the German Bight they are known mainly from the island of Heligoland. The species is also known from the Azores and Canary Islands. Due to this distribution area it does not seem to be a particularly cold-loving species. Species like this benefit from two phenomena. On the one hand, the warming up of the North Sea water causes an ever-increasing advance northward. On the other hand, however, such species are increasingly finding a foothold on the pillars of offshore wind turbines. And also on the "rubbish reefs" on the seabed of the North Sea, where they would otherwise not have found a foothold on sandy substrates. This species also keeps well in the aquarium at room temperature and behaves peacefully and not aggressively towards other animals. Therefore they are ideal and colourful aquarium animals - if you can get them alive. They are also long lasting and durable. The strigose squat lobster prefers to position himself into the flow to lurk for small food particles. The specimen shown on the top right sits near the aquarium pump and waits for food particles to drift towards it. So far, such crabs have only been caught very rarely by our shrimp trawlers. But this could change very soon during the next decade...

Dwarf Hermit Crab, *Diogenes pugilator* (Roux, 1829)

The **Dwarf Hermit Crab** is regularly known from the Mediterranean Sea unto the Dutch coast. It reaches a carapace length of about one centimetre and therefore only inhabits snail shells up to the size of a periwinkle shell. I was very surprised when in August 2008 I received a small hermit crab from a shrimp trawler in Neuharlingersiel, which was unknown to me from our coasts until then. For the first time I had made acquaintance with these animals through shellfish shipments from the Mediterranean Sea. And also during my Mediterranean holidays I had found this hermit crab every now and then. But now, due to the warming up of the North Sea water, this species seems to advance further and further north. On Norderney and Baltrum the dwarf hermit crab can already be found regularly in the waddens while the summer season. And since the warm winters starting in 2015, this species was already collected in March (!) by the shrimp trawlers off the islands Juist and Norderney as by-catch. For the sake of completeness it should be noted that these animals can often be found among nylon bundles, rubbish and netting remains. They also like to bury themselves in the sand next to the common true shrimps, which is why they are often part of the fishermen's by-catch. At room temperature they can be kept very well. Due to that they must be taken as typical indicator-species for the warming up processes in the North Sea.

 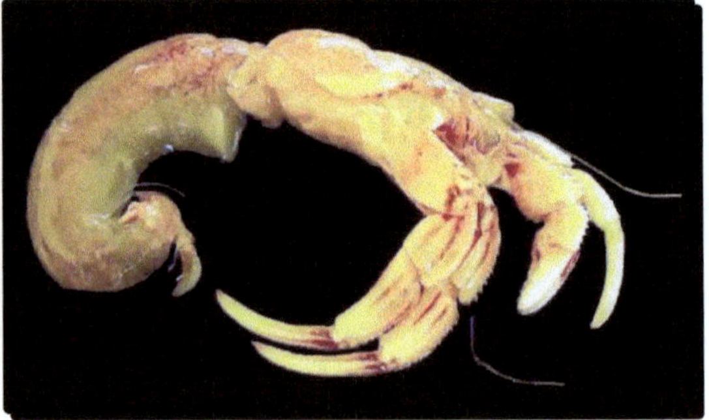

The **Common Hermit Crab** lives in snail shells to protect its soft abdomen against predators. Hermits can reach a body length of about 10 centimetres. Sometimes, at low tide, large numbers of small hermit crabs can be found in the tidal flats and puddles in the waddens. They are only about one centimetre in size and live mostly in the shells of periwinkles. With a bit of luck, you can also find slightly larger hermits in the tidal flats of the islands, which then already do carry the shells of moon snail or whelks. Hermit crabs shed their skin like any other crustaceans, but have a special problem. Because their snail shells do not grow with them, they have to move into a larger one. Since larger shells are not available near the shore, larger hermit crabs are usually found below the tidal mark. Juvenile hermits are mainly found near the coast from April to September; in winter they move to deeper areas. Hermits are easy to keep in aquaria and require at least a medium-sized whelk shell after only one year. Hermit crabs voluntarily leave their shell only, when they suffer from lack of oxygen or when they have to move. For a short period of time they can tolerate higher water temperatures in summer, especially when they sit in tidal pools during the ebb tide. In the long and medium term, however, they need cold and oxygen-rich water again and again in order to survive. They will most likely benefit from a general warming up of the North Sea water as long as there are snail species under these conditions, whose shells they can inhabit. If the snails disappear the common hermit crabs will follow them just a little bit later.

Since the early 2000s, at least two newly introduced species of this crustacean family have been found in the German Bight. In the meantime, they have already conquered the coasts of Schleswig-Holstein as their new habitat. Originally, they were brought by ships from Japan and Korea to the cargo port of Rotterdam, from where they continued their triumphal march to Northern Europe. They seem to require brackish rather than seawater for successful reproduction, but this does not seem to be a problem for these species due to our topographical conditions. If you want to detect them at a location, it is usually sufficient to turn over a few loose stones during the ebb-tide. Here they sometimes sit in large flocks or even sporadically to wait for the next flood. Like almost all crabs of the littoral zone, they are frugal omnivores, but they have a special preference for certain types of algae. Among themselves they are very peaceful and show a positive social behaviour, which might be a reason for their spreading success. In addition, they should also benefit from the warming up of the North Sea, as warmer water also means more plankton food for their larvae to a certain extent. Whether they will displace our native crab species is not yet clear. But it is a fact, that they still do share the habitat. And often both species can be found together under the same stones...

Japanese Shore Crab
Hemigrapsus penicillatus
Approx. 30 mm wide

Pacific Shore Crab
Hemigrapsus sanguineus
Approx. 40 mm wide

 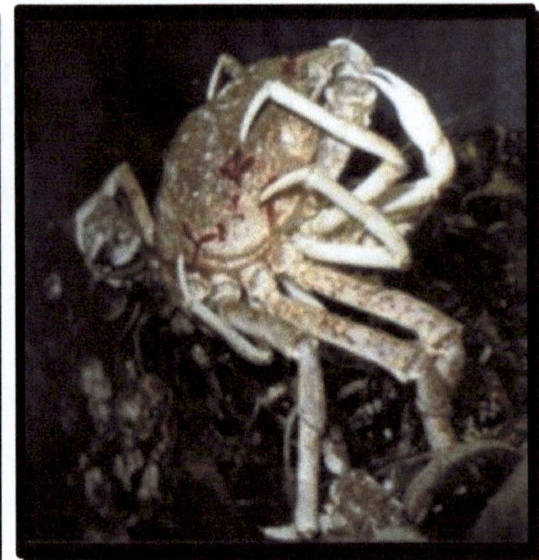

The **Great Spider Crab** can reach a leg span of about 30 centimetres. It can be found on hard substrates rich in structure with algae growth from a depth of about 10 metres. In contrast to other crabs, which flee from enemies quickly, the spider crab has developed a different strategy. The spider crab camouflages itself with algae, sponges and hydroid polyps, which it does collect from the sea floor. Sometimes they even use small nettle actiniae. They attach this material to their carapace with the help of body secretions. This is how the spider crab deceives its enemies. It also moves very slowly so that it can hardly be noticed. After the winter spider crabs mate at a temperature of 10°Celsius. In the copula (top right) both partners sit opposite each other in the vertical and claw into each other. The sexual intercourse can last several hours before the partners separate from each other. A few days later, the female then ejects the eggs into her brood pouch located under the body, where they are fertilized by the male's sperm supply. The great spider crab clearly belongs to the subarctic faunal circle and cannot adapt to higher temperatures. Since 2014, this crab was caught only very rarely by shrimp fishermen as a by-catch. It will probably very soon be extinct in the southern North Sea; I presume during the running decade...

The **Edible Crab** can reach a width of up to 30 centimetres without legs. This makes it the largest and heaviest crab species in the German Wadden Sea. Edible crabs are vicious predators, which make prey of anything else, that is not able to flee. Since they are not very fast, especially mussels and snails are ideal prey. In addition, the webs of blue mussels offer the juvenile edible crabs ideal cover. Conversely, adult edible crabs sometimes carry commensal sea anemones and other epidemic organisms on their carapace, which benefit from the mobility and the fine food remains of the crab. Edible crabs have a distinct sexual dimorphism, with males having significantly larger claw legs than females. The females also have a slightly rounder carapace than the males. They reproduce similarly to the **Shore Crab *Carcinus maenas*** and can produce up to a million eggs. Recent research has shown, that edible crabs can even displace the **Lobster *Homarus gammarus*** from its burrows, and that their population near Heligoland is increasing, while that of the lobster has decreased. They also benefit from the general warming of seawater, as they are not necessarily among the cold-loving species. Because the edible crab is also known from parts of the northern Adriatic Sea in the Mediterranean. However, it is not certain whether the species was introduced there by human activities, or whether it has always been native there. Small crabs of only one to three centimetres in size seem to be durable for several weeks to months, even at room temperatures of about 20° Celsius, according to my experience in keeping them in marine aquaria. However, the premise is that the animals should be accustomed to these temperature ranges as slowly as possible. Should other crab species or large crabs such as the lobster give way to climate change, it can already be considered very likely that the edible crabs will successfully conquer their ecological niches. Furthermore, warmer water in this case also means faster growth for the broods of the edible crab, so that it could even take over a dominant position in the ecosystem of the German Wadden Sea in the closer future.

Juvenile Edible Crab, Approx. 50 mm wide.

A female Edible Crab may produce unto one Million Eggs. A good mating strategy for a successful species!

The tribe of the echinoderms is very diverse. The classes of sea urchins, starfish, brittle stars, sea cucumbers and feather stars belong to this tribe. The echinoderms only occur in salt water, which could be due to their larval development. They rarely tolerate low salinity levels, and relatively few species occur in areas subject to strong tidal influence. All in all, they occur from shallow water to the deep sea and have occupied various ecological niches. They are also very variable within their respective classes. For example, there are species of sea urchins that scrape algae from the rocks in shallow water, species that live buried in mud and only eat detritus, and others, that live in lightless depths and use carrion and other particles as food. It has been scientifically proven that some echinoderms, such as the **Common Starfish Asterias rubens**, have photoreceptors in their epidermis (skin), with which they can distinguish light from dark. All echinoderms share an endoskeleton consisting of small calcareous platelets, which are covered with spines. The entire body of the echinoderms is traversed by a hydraulically working system of small vascular pathways, which is called the ambulacral system. Almost all echinoderms (with the exception of the class *Crinoidea*, the sea lilies and feather stars) have small suction feet which are connected to this system of vessels with which they can attach themselves to the substrate. In doing so, they develop such powerful adhesive forces, that they are able to hold on in the tidal zone or in areas with strong currents and, in addition, to hold and to open prey. In some species, but especially in tropical and subtropical sea urchins, some of these so-called pedicellaria are transformed into small poison tongs, with which the sea urchin can pinch and nettle sensitively. The reproduction of the echinoderms follows different strategies, so that it is very difficult to make sweeping statements here. Most echinoderms reproduce by ejection of sperm and eggs into the open water, whereby the populations of one species usually spawn synchronously. The animals orient themselves by the phases of the moon. In tropical coral reefs they even spawn synchronously with many corals.

Below are therefore some echinoderms' larvae:

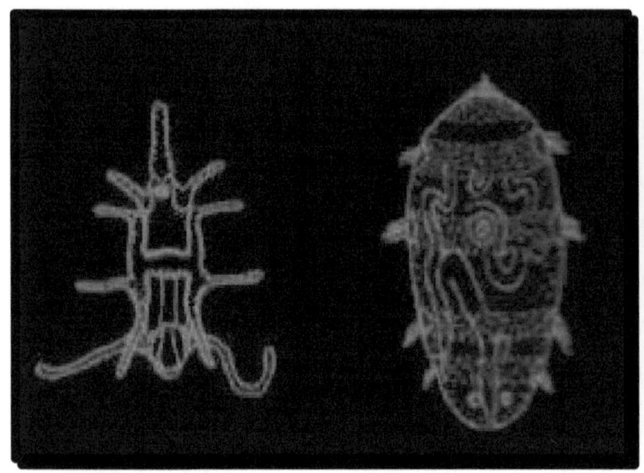

Starfish larvae.

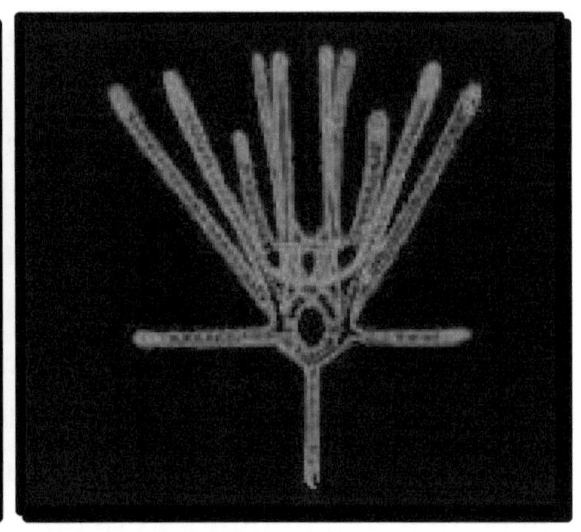

Brittle starfish larva.

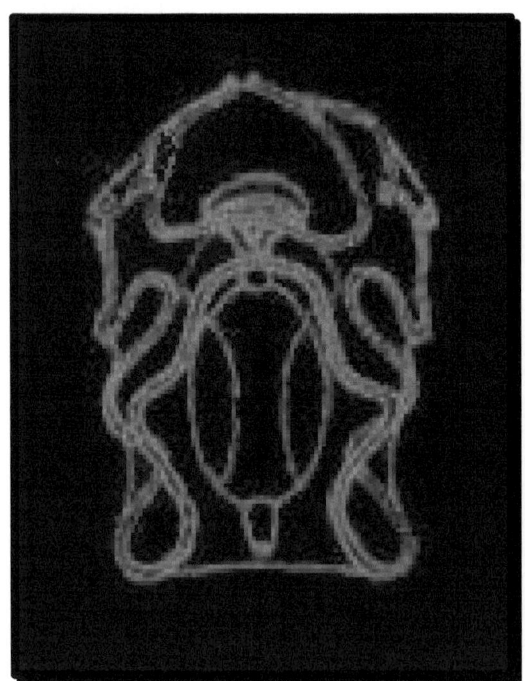

Sea Cucumber larva.

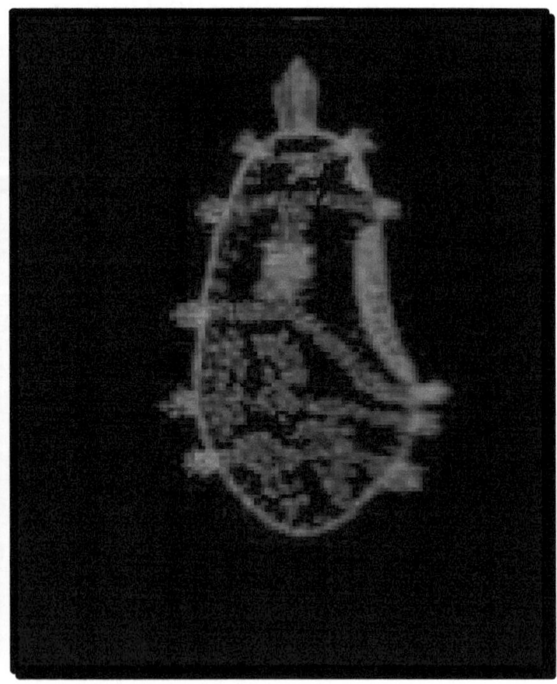

Feather Star larva.

In addition, some sea cucumbers and starfish still use the reproduction strategy of vegetative division, whereby one arm is tied off to form a new animal, or the animal is divided into two halves. Such species are also known as potentially immortal, as their descendants may still have some of their original tissue in their bodies. In addition, some species, such as the **Blood Star *Henricia sanguinolenta*,** have been shown to actively care for their brood by guarding their eggs until the juveniles' hatch. Unfortunately, man has exerted a disastrous influence on the fauna of the North Sea, as evidenced by the fact, that real rubbish reefs have formed on the bottom of the North Sea. These consist of a mixture of old nets, plastic and industrial waste of all kinds, as well as algae, sponges and bryozoans that have settled on them. And juvenile echinoderms, such as the **Common Starfish *Asterias rubens*** or the **Green Sea Urchin *Psammechinus miliaris*,** also live in this chaos. I found the latter several times, preferably on old plastic barrels with vegetation, when a shrimp trawler was delivering the garbage in the harbour. Smaller specimens of this sea urchin even survive several hours on dry land and can even recover from the catch. The same applies to small starfish up to a size of about 5 centimetres. This is how chemical substances and toxins get into the marine food chains via the invertebrates settling on the garbage, at the end of which humans are also found. Finally, it should be noted that some echinoderms are also good indicator organisms, which tell us, how intact our environment really is. For example, even on the East Frisian islands, the green sea urchins, which used to occur there in the past, can now only be found very rarely or not at all. For this reason, research trawlers regularly collect and examine such species at very precisely defined measuring points. And in muddy places such as Norddeich one unfortunately searches for starfish and sea urchins in vain. And also brittle stars and sea potatoes cannot be found here. Another questionable fact is the absence of starfish, which in earlier times were often found in association with the blue mussels. Nowadays these are only found - if at all - on the offshore islands. If at all... Could it be that these species have developed a "heat problem"?

Common Starfish, *Asterias rubens* Linnaeus, 1758

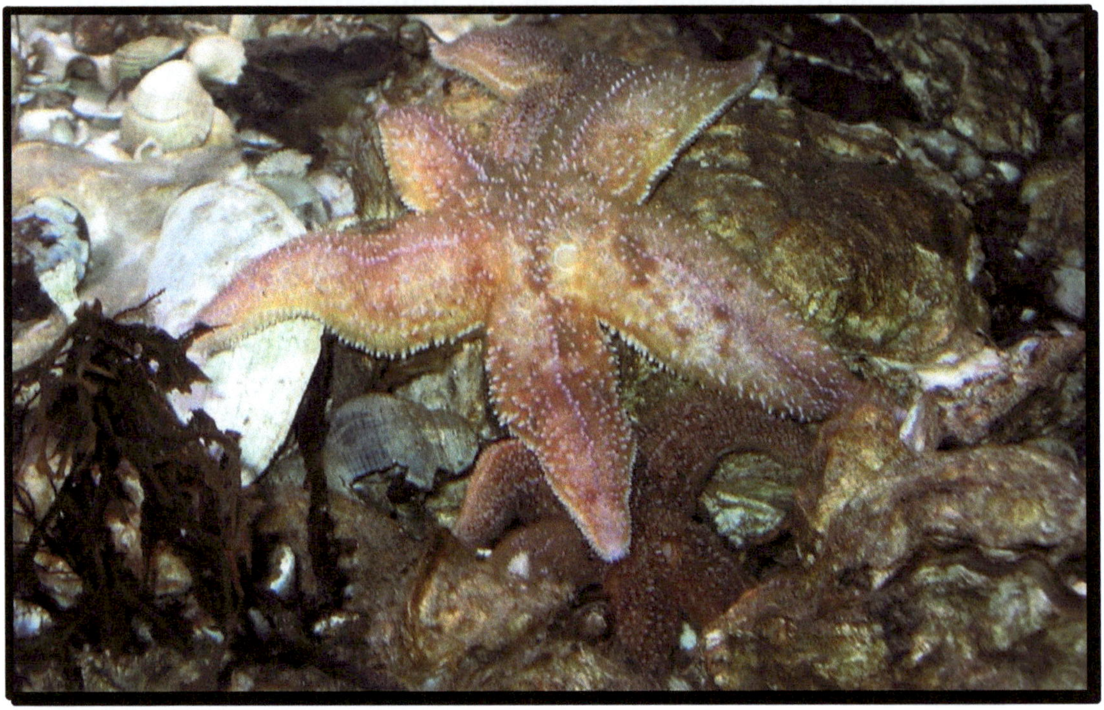

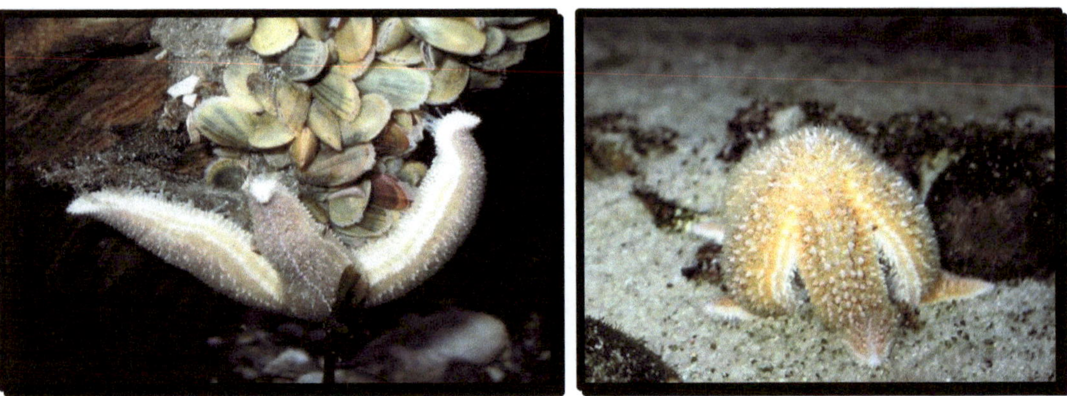

Starfish in large aggregations have become a real exception on the coasts of the German North Sea.

These starfish make prey of blue mussels, their preferred prey.

The **Common Starfish** can reach a maximum diameter of 40 centimetres. Smaller starfish already occur in the tidal zone, where they should actually be found in large flocks on groynes and stones with mussel growth. In the meantime, however, starfish have become a seldom species almost everywhere. The common starfish is a predator, that mainly feeds on blue mussels, which he does embrace and slowly pulls apart. This process can take several hours, but the starfish mainly takes advantage of that the mussel has to breathe fresh water at some point. As soon as it has opened its shells a tiny piece, the starfish puts his stomach between the mussel shells, injects a stomach secretion into it and then digests the prey outside his body. Then he sucks in the mash. Starfish also eat carrion and anything they can overcome. Since the heat year 2016, there has been a hidden starfish extinction in the southern North Sea, which has hardly been noticed so far. However, I know from a reliable source, that on the groynes of the North Sea island Borkum, where you could usually find many starfish associated to stones during the ebb-tide at any time of the year, suddenly there were only a few very small juveniles left. And also in the North Sea aquarium Borkum the starfish plague spread and killed the stock. Until then, this phenomenon was well known from the Pacific coasts of North and Central America, where millions of starfish died from Alaska in the north to Baja California in the south. This was a great starfish disease caused by a virus. This illness is caused by a denso-virus in connection with increasing water temperatures. The starfish get spots on their skin and their connective tissue dissolves. The animals then die within a few days. The viral disease is contagious, so that stocked animals die particularly quickly in aquaria in a confined space. Apparently, this decline is more likely to take place in the shallow shore areas that have become too warm, because the shrimp trawlers still catch many starfish offshore. Therefore the problem seems to be limited. But the warming up of the shallow shores of the southern North Sea during the last two decades could be one of the main reasons for the disappearance and permanent absence of the common starfish.

The **Green Sea Urchin** is regularly found on hard substrates, on waste accumulations or on driftwood. It reaches sizes of about 5 centimetres in diameter. Where sea urchins have crawled over a stone, they leave clear feeding marks, because together with the algae they also eat stone particles, which they need to form spines and endo-skeletons. The green sea urchin also eats driftwood, seaweed and other algae, although it is not found on soft or muddy soils in the tidal zone. Until the 1980s, it was already found in the shallow waters near the shore in the southern North Sea, especially the East Frisian Islands. This was particularly true of the groynes on the island Borkum. However, this sea urchin has been absent there for at least the last 3 decades. In the southern North Sea, green sea urchins can no longer be found in the tidal zone nowadays. This could be due to the contamination of their substrates with various pollutants, but could also be due to the excessive warming up of the shallow tidal areas in the Wadden Sea.

The **Edible Sea Urchin** lives on European rocky coasts and can be found from a depth of about 10 meters. It can reach a diameter of up to 16 centimetres, and its spines are about 2 centimetres long. With its strong teeth, the **"Lantern of Aristotle"**, this sea urchin eats both seaweed and algae coverings on rocks. It also eats carrion and small ground animals. On the island of Heligoland there is still a reproductive population of this cold-water species, which has been placed under strict protection. Too high temperatures beyond 16° Celsius are not tolerated by this species for long. This means, that this sea urchin could quickly become a victim of climate change. Then at some point you can read the laconic note on the Internet: **"Edible Sea Urchin (*Echinus esculentus*)**, extinct in the German Bight since the warm index year 20XY".

Endo-Skeleton.

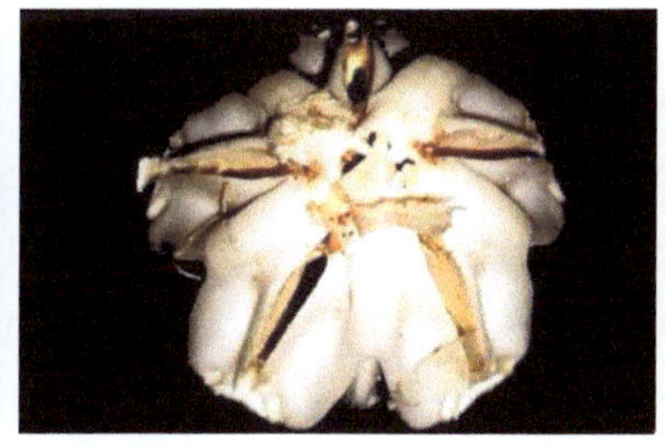

Lantern of Aristotle.

Single teeth.

The **Spiny Starfish** is the largest starfish in the North Sea and he may reach a radius of up to 80 centimetres. He is a widespread species, because he occurs from the Mediterranean to the Atlantic Ocean. His German name "ice-starfish" does not refer to the temperature ranges he prefers, but to his colouring, which can also be whitish. He is a strong predator. The spiny starfish tolerates higher water temperatures of up to 18° Celsius, which is due to his wide distribution as far as the Mediterranean. However, in the Mediterranean he does not penetrate into the shallow shore zones, where considerably higher temperatures can prevail. That is why he can only be found there from about 10 metres depth. Since about 2003 I occasionally receive by-catches from shrimp trawlers of the southern North Sea, but they have never had a specimen in their by-catch. Is this because it prefers hard substrates, or has the southern North Sea already become too "warm" for it?

From a purely taxonomic point of view **Tunicates** belong to the vertebrates. All **Sea Squirts** are considered to be part of the subphylum of the **Tunicates** together with the **Tailed Tunicates** and the **Salps**. Since their larvae have a chorda, i.e. an organ similar to the vertebral column, some biologists regard these animals as precursors of vertebrates. But they do ignore the fact, that some species also reproduce by forming side buds. Originally, Aristotle tried to divide the animal kingdom into vertebrates and invertebrates more than 2300 years ago, but it is a fact, that such a blanket division of the animal kingdom causes some problems. Because there are animals in basically intermediate forms that cannot be assigned to one of these two categories easily. Sea squirts are one of these problematic groups. On the one hand, their larvae owe a chorda, which is a vertebral column like organ. On the other hand, they show no other characteristics of other typical vertebrates such as limbs or a permanently present brain. Some sea squirts, however, have a heart-like organ and associated pathways. It would not be surprising, therefore, if the taxonomic position of these animals were to be completely revised one day. Sea squirts live as filter feeders, whereby they differ from sponges in that they always have one inlet and one outlet. Sea squirts are sometimes found in the shallow waters of the tidal zone. They are found especially on mussel shells and other hard substrates such as plastic waste. In contrast to sponges, sea squirts seem to be less sensitive to air, because they can contract themselves, if necessary, to prevent dehydration. Due to international shipping, many species have spread all over the globe, which can make an exact determination very difficult. In some cases they have become a plague, that mainly infests ships' hulls and ports, which then results in expensive cleaning measures. Sea squirts are only suitable as climate indicators to a very limited extent, because their distribution is often of anthropogenic origin and they have a sedentary lifestyle. On the other hand, the seasonal occurrence of their zooids allows conclusions to be drawn about the current course of a reproductive season, and thus indirectly about changing climates.

The **Yellow Sea Squirt** can reach a length up to 15 centimetres. It lives solitary, but can occasionally thrive in small colonies. Particularly characteristic of this species are the yellow rings around its siphonal openings and its long and slender body shape. Yellow sea squirts look gelatinous and transparent, but often delicate shades of green shimmer through their tissue. Nowadays this species has been spread worldwide. And it is one of the pioneers that suddenly appears out of nowhere, if it finds optimal conditions. They also colonise artificial substrates such as buoys, harbour walls, plastic waste or ropes floating in the water. Although these sea squirts do not belong to the particularly long-lived species, once they have conquered an area for themselves, they are continuously present there. Under natural conditions, the vase sea squirt reproduces sexually from June to September, producing both eggs and sperm. The occurrence of their larvae in marine plankton and their adult zooids can be related to the course of the seasons, which can then provide indirect conclusions about the warming up of the North Sea.

Star Squirt, *Botryllus schlosseri* Pallas, 1766

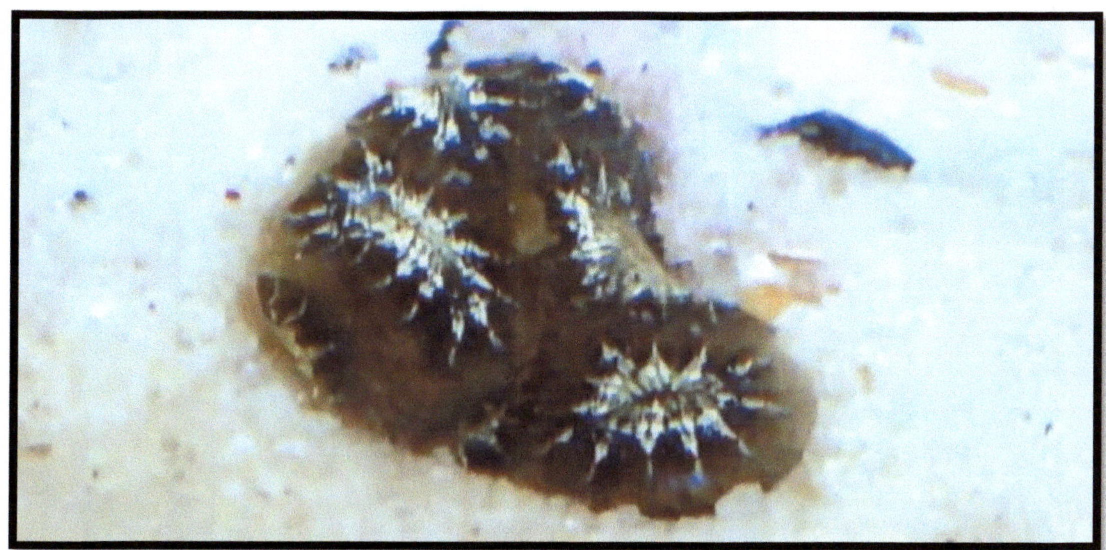

The **Star Squirt** is spread from the Mediterranean Sea around the British Isles to Norway in the north. Yet this species has probably become a cosmopolitan, having been spread by ships as far as North America, Iceland and even the Pacific Ocean. This sea squirt forms small gelatinous colonies. These individual zooids of a colony reach a radius of about 10 millimetres and are always grouped in a star shape around a common outflow opening. They are usually bluish or even purple. The star-shaped pattern on the single zooids makes this species unmistakable. They are found on swimming pontoons, groynes and other hard substrates in shallow water. Here the star squirts unfold their life cycle, whereby the adult colonies dissolve in the winter and then appear in spring as if from nowhere. The star squirt can reproduce both by forming buds and by sexual reproduction with the help of eggs and sperm. The larvae of this species can be found in plankton from May to October. It may be assumed that in the southern North Sea, due to the lack of the "winter" season, it will be found earlier and earlier in the plankton and its settlement substrates. At least since the year 2016, their adult zooids had already been present in the harbour of Norddeich in March and April.

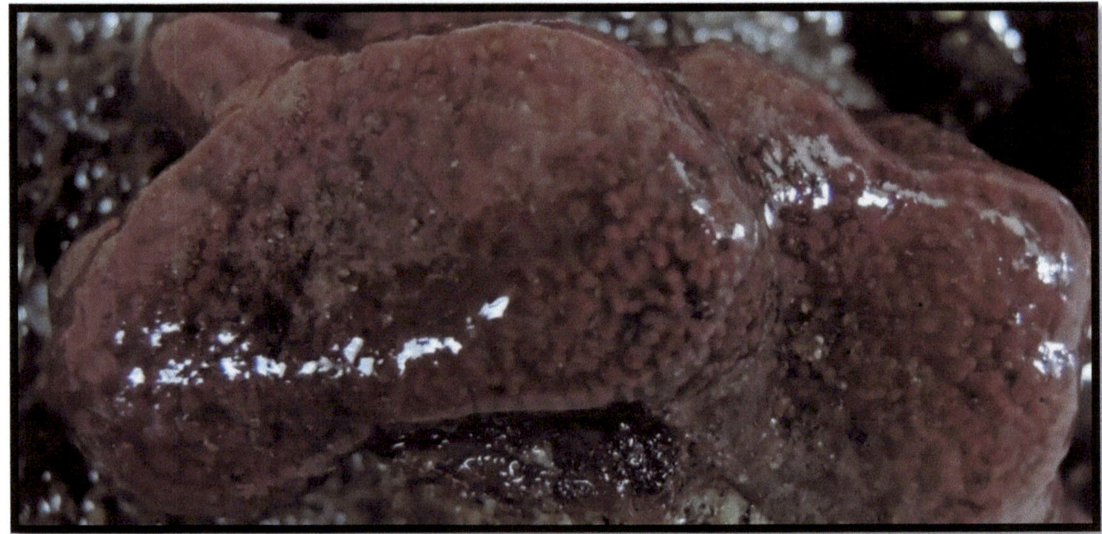

The **Botrylloid** can be found in harbours of the southern North Sea, especially at the floating docks of small boats. In addition, the botrylloid was spread all over the world in the 20th century. It mainly colonises hard substrates from shallow waters unto a depth of 1000 metres, which is an amazing distributional pattern for a "shallow water species". The special characteristic of this species are the zooids which are always arranged in double rows, each of which is about 2 millimetres high. Their colour can be pink, yellowish or beige. They are always covered with a layer of jelly. And the larger the colony becomes, the less well the characteristic double rows can be seen. In spring, the first zooids begin to colonize suitable substrates, but the young colonies maybe so small that they can hardly be noticed. But from June they mean colourful spots on oysters or floating objects such as buoys and fenders. They reach their growth peak in September and October. During the winter the colonies shrink and become regressive. They store their substance in so-called opaque ampoules, from which new colonies are formed in the following spring. The time of their "resurrection" can provide conclusions about the increased sea temperature. In 2020, the first small colonies of this species were already visible in April and May.

Folded Sea Squirt, *Styela clava* Herdmann, 1881

The **Folded Sea Squirt** reaches a size of about 10 centimetres and usually occurs in pairs or in small groups. It can contract at low tide to prevent possible dehydration. The folded sea squirt has migrated into our waters from the North Pacific, i.e. from the coasts of Japan and Korea. In the meantime, it can be regularly discovered on the East Frisian Islands and also in protected harbours on the mainland. They settle on hard substrates and prefer areas where they cannot dry out during the ebb-tide. However, they do not mind if they`ve been put on dry land. It has been proven, that they can survive this for up to 48 hours, even when the sun is shining! This attitude also explains their now global distribution by ships. From about April onwards, the zooids of the folded sea squirt can be found on the floating pontoons of our harbours. Here they grow quickly to their final size. In October and November they disappear completely and reappear in the next spring. As the folded sea squirt reproduces itself in our waters without any problems, it has obviously found optimal warm climatic conditions in the German Bight, which makes it forget its origin in exotic regions...

All sharks and rays have been systematically classified as members into the **class *Elasmobranchii*,** i.e. cartilaginous fish with gill-plates. This is because the gills of sharks and rays are connected to cartilage-like plates, which distinguish them anatomically from all bony fish. Sharks and rays often have splash holes through which they suck in their respiratory water to squeeze it out through the gill slits. These splash holes are found in bottom-dwelling as in free-swimming species also. In the case of bottom-dwelling species, these holes prevent sand from entering the gills, as the animals, lying on the ground, suck in their respiratory water from above and press it out again downwards. In the case of pelagic species, the splash holes ensure, that the shark can draw enough oxygen from the water to breathe even when eating prey or swimming very fast. Some shark species are demonstrably no longer alternating warm animals, as they are able to generate a higher body temperature than the surrounding water through muscle contractions. This gives them an enormous advantage over prey, that is not as fast as the shark. On the other hand, such predators have an increased energy requirement, that requires a relatively large number of prey animals. Sharks and rays do not have a skeleton, but only a spine made of cartilage. In addition, they usually have very pronounced jaw bones, which, however, are not firmly attached to the skull or the teeth. The teeth sit on a cartilage strip, where they are arranged in several rows behind each other. These teeth are therefore also called "revolver teeth". If a tooth breaks off, it falls out and a new tooth automatically slides in from the next row. Furthermore, the skin of sharks and rays is covered with small skin teeth. These are all oriented towards the back and improve the hydrodynamics of the animals considerably. If you want to stroke a shark or ray, you will find that because of the skin teeth, this is only possible in one direction, namely towards the tail. In the other direction you get stuck. Shark skins used to be processed into sandpaper or raincoats. Some nasty injuries, which divers have suffered when dealing with sharks, have only been abrasions! Sharks and rays possess a special sensory organ, namely the ***Lorenzini ampullae***. These are small channels filled with mucus, which are usually located around the snout in the head area of the animal and are connected to its` brain. With this organ these animals can locate even the smallest electrical fields in the water. Because every living being has its own small electrical field, cartilaginous fish can use this electrical receptor as a prey radar. Most shark species in the North Sea are harmless to humans, but spiny dogfish have poisonous spines in their dorsal fins. One of the more dangerous shark species is the **Porbeagle *Lamna nasus*** (pictured below), which belongs to the same shark family as the well-known great white shark and hunts for fish in the North Sea. Although this species has so far only been known to have incidents with bitten fishermen,

it has a remarkable set of teeth and sometimes swims into fresh water, where it migrates upstream. Some dangerous tropical shark species are found hundreds of kilometres upriver in inland areas, where they can actually be dangerous  for bathers. In this respect, the porbeagle should give you pause for thought, as a specimen of this species was actually pulled out of the Eider river by anglers a few years ago... Among the rays, we should mention the **Stingray *Dasyatis pastinaca***, which occasionally appears in the North Sea and has barbed spines on its tail stalk, and the **Marbled Electric Ray *Torpedo marmorata***, which is capable of delivering electric shocks. With a constantly rising water temperature in the North Sea, the occurrence of subtropical shark and ray species in the German Bight is therefore to be expected to increase in the future. Be it that these animals only follow their prey to the north, or be it, that they now find ever better living conditions for their own offspring here.

The **Spiny Dogfish** is a cosmopolitan that can be found in the North Atlantic and North Pacific as well as in the South Atlantic or southeast of Australia. It is probably the most common shark species worldwide. The animals live in large shoals of thousands of animals and hunt for cod, herring, other swarm fish and bottom-dwelling invertebrates. Females can reach 1.20 metres in length, males about 90 centimetres. Dogfish can live up to 60 years and reach sexual maturity in the male sex at around 10 years of age, while the female reaches maturity at 12 years. Spiny dogfish are usually caught at depths between 10 and 200 metres as a by-catch in cod fishing. Nowadays, spiny dogfish are also commercially marketed. Their raw belly lobes are offered as "sea eel". Their low reproduction rate cannot compensate the amount of caught spiny dogfish, so it is already foreseeable, that one day the spiny dogfish will have to be put under protection because of this predation. If one day the spiny dogfish should disappear completely from the German Bight, however, this will probably not only be due to overfishing but to climate change. This is because spiny dogfish migrate to greater depths if the surface water becomes too warm for them. Or they may retreat to the more northern parts of their range. Dogfish are extremely rare by-catches in shrimp-fisheries. I personally only know of one single catch from the shrimp fishermen in the southern North Sea in the period 2012 to 2020, and this was a new born juvenile in spring 2016. If the warming trend continues in the southern North Sea, it might have been the last specimen here.

Dogfish or Small Spotted Catshark, *Scyliorhinus canicula* (Linnaeus, 1758)

The **Dogfish** is the most common shark species on the European coasts. It occurs both in the algae zone and over sand and scree bottoms, and has been found at depths ranging from 10 metres to 780 metres. The maximum length of these sharks is about 1.20 metres and they can reach an age of at least 9 years. Catsharks are very well suited for being kept in aquaria, as they do not require as much space as free-swimming shark species. As these animals can reach lengths well over 50 centimetres even in an aquarium, they must nevertheless be kept in suitably large aquaria. They feed on squids, shrimps, worms, small crabs and small fish. When they are not hunting, they doze lazily on the bottom of the aquarium, thus saving their energy. Catsharks prefer to be active when feeding, then they turn into elegant swimmers. Catsharks are occasionally caught by sea anglers, and they are sometimes a by-catch of commercial fishing. Their meat has a good taste, but always tastes a little bit like ammonia. Catsharks have an excellent sense of smell and are particularly good at attracting squid pieces. In a suitably large aquarium, several sharks can also be accommodated. Under good living conditions, they often proceed to reproduction. Catsharks mate, in which the male wraps his very flexible body around the female and fertilizes her internally. A few weeks after mating, the female usually lays 2 egg capsules on each kelp stem, circling the kelp so that the eggs can properly anchor themselves to the kelp with their spiral-shaped adhesive threads. These egg capsules are also known as mermaid`s purses. A female catshark can lay a total of 18-20 eggs during one mating season. The development time of the eggs depends on the water temperature and usually takes 8-10 months. At the beginning you can see the egg yolk in a freshly laid egg capsule of the catshark in the middle. If the egg has not been fertilized, the yolk dissolves into a mushy mass after a few days. If the egg has been fertilized, one can study the development of an embryo until it develops into a small shark with yolk sac. The sharks only hatch out of the egg when the embryonic yolk sac has been consumed. This phase of shark breeding is the most delicate, because the small sharks must now be brought to suitable food. If they refuse to take food, they are condemned to starvation. The offspring of the small spotted catshark has often been successful and is therefore often shown in public aquaria. Catsharks have very beautiful golden eyes, which they can close with an eyelid. This eyelid distinguishes them from all other known fish species. Dead catsharks therefore often have closed eyes. Close behind their golden eye catsharks have a splash hole, through which they press the breathing water through their 5 slits of plate gills. This splashing hole is a typical adaptation to bottom life, with which the animals avoid inadvertently inhaling sediment particles on sand or muddy ground. When swimming or resting slightly raised on the bottom, they can of course also inhale through their mouth. Catsharks can also waddle across the seabed on their two large pectoral fins, providing a cute sight. This is often further enhanced by the fact that they move in a snake-like fashion. In recent years, catsharks of various sizes and egg capsules have been caught by shrimp trawlers in the southern North Sea. Unfortunately catsharks like to lay their eggs on old fishing nets and other rubbish. And they also seem to benefit significantly from a warming up of the North Sea, because at higher temperatures their eggs can develop faster. So they have a clear advantage over other species both from littering and climate change, what makes the prognosis for the further existence of their species in the German Bight good for any case.

Tope Shark or Sweet William, *Galeorhinus galeus* (Linnaeus, 1758)

Tope Sharks can reach a total length of about 2 meters. They belong to the subtropical species that reproduce in the southern North Sea in summer. Here they give birth to their juveniles off our islands. Tope sharks are also regularly found in subtropical waters around the world, such as off the Brazilian coast. Starting in 2014 and in the following years, juveniles and in some rare cases even adult specimen were caught off the East Frisian Islands in May and June. Obviously, these subtropical sharks reproduce successfully here. The females of this species give birth to alive young sharks, which can be between 35 and 50 centimetres long. If the North Sea warms up even more in the foreseeable future, such catches and sightings will certainly increase. Tope sharks only grow to just under two metres in length and feed on small fish and crabs, so they are not dangerous to humans. They represent an enrichment of the ecosystem in the southern North Sea and should therefore be protected as much as possible. This is because sharks regulate the stocks of fish, but also of cephalopods, and thus help to keep natural food chains and the overall ecological structure of the North Sea intact. Tope sharks do benefit very well from the climate change and the number of caught specimens may be a further prove for this theory.

The **Electric Marbled Ray** is actually more familiar from the Mediterranean and the Eastern Atlantic, where it can reach sizes of about 60 centimetres. Electric rays have muscle plates that are superimposed in several layers in their pectoral fins. They rub against each other and generate electric shocks to paralyse their prey. During the day, they lie buried in the sandy bottom, while they go hunting at dusk and at night. They can therefore become dangerous for divers, who make night dives. Electric rays are viviparous fish and give birth to between 5 and 30 juveniles after a pregnancy of about 10 months. The specimen shown here was caught 2004 in the Baltic Sea and then exhibited in the Multimar-Wattforum Tönning. Unfortunately the animal could only be kept alive for a few weeks in the aquarium. Whether this specimen can be seen as a proof of the continuing global warming up, or whether it must be considered as a rare maddening guest, is unclear so far.

The **Broadnosed Skate** actually occurs regularly in the Mediterranean Sea, on the North African coasts and in the English Channel. In spring 2015 and in the following years including May 2020 several specimens were caught off the East Frisian Islands. This can be seen as further evidence of the rapidly progressing global warming up, as the winters since 2014 have been mostly too warm. On the coast of Lower Saxony, there were hardly any such weather events such as snowfall or severe frost. The broadnosed skate can easily be distinguished from the **Thornback Ray *Raja clavata*** by its` numerous small black spots on the upper side of its body. Furthermore, the males of the broadnosed skate have a row of spines only in the middle of the body, while the back of a male thornback ray is covered with several rows of thorns. In addition to that broadnosed skates always have some larger beige spots on their back. They can reach a body length of 120 centimetres and are present from shallow water to a depth of about 400 metres. They are quite robust fish and usually survive the catch by a shrimp trawler. As the largest catches from 2015 onwards were females and juveniles, it can be assumed that they came into the German Bight to lay their eggs. Because the southern North Sea now seems to have the right temperature and the right food supply for their offspring.

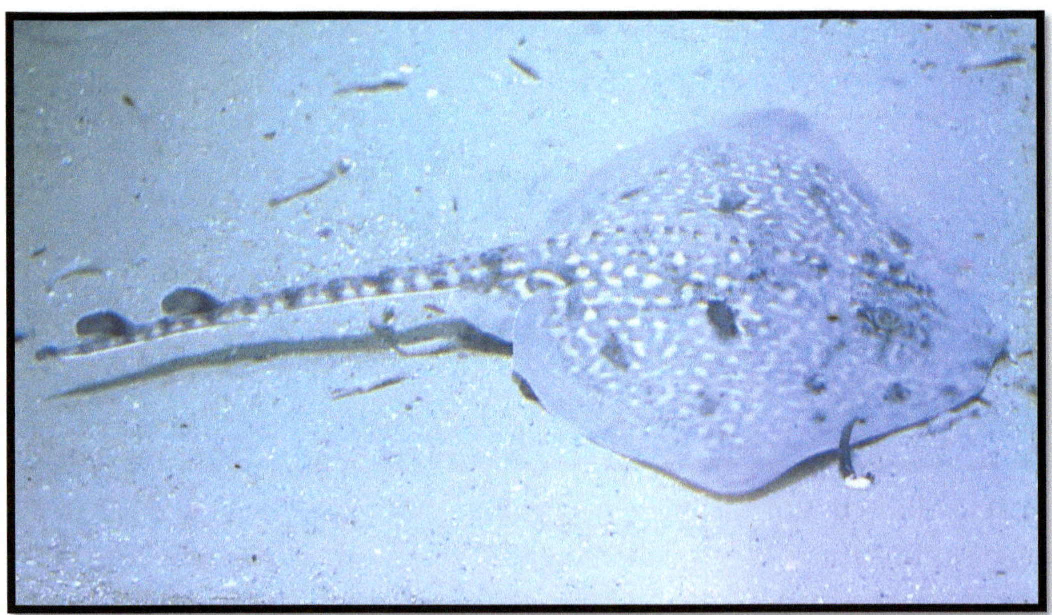

The **Thornback Ray** has numerous spines not only on the upper side of his body, but also on the underside. He is a cartilaginous fish that has such a wide range of distribution that he might be a cosmopolitan. He occurs from Iceland in the north unto the Mediterranean and the Black Sea. As well as to the South African coasts in the south. This species can reach a length of about 120 centimetres. Even though the population of this ray is declining off the German coasts, and its typical black egg capsules with the four "horns" are increasingly rarely found in the wadden areas, this species is not threatened with extinction because of its large distribution and its relatively high reproduction rate. This is because the thornback ray can reproduce at a body length of about 70 centimetres, and in just one season a female thornback ray can lay 70-150 egg capsules. Which means a very high reproduction rate for a cartilaginous fish. The juveniles hatch out of the egg capsules after 4-5 months and have a total length of about 12 centimetres. Whether this species will reappear more frequently on German coasts in the future, or will be replaced by the broadnosed skate, is not clear yet.

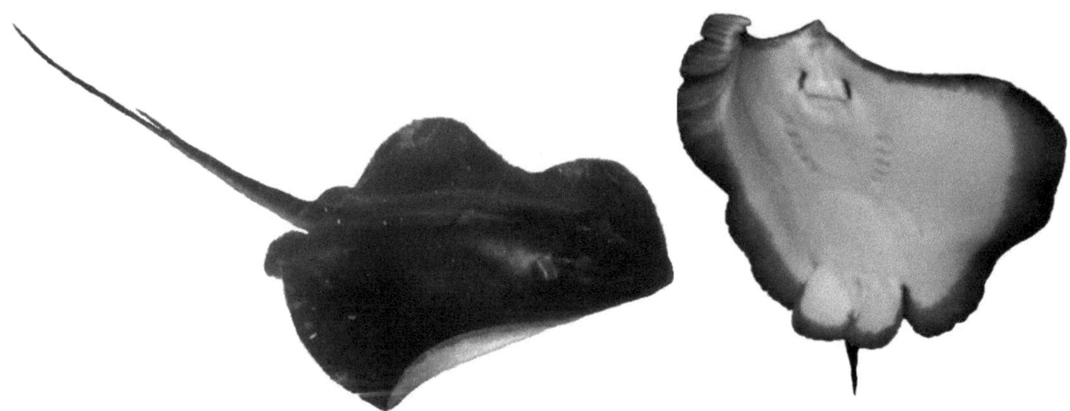

The **Stingray** has a very large distribution area and is found from southern Africa, up the East Atlantic coast, around England, Ireland and in the Mediterranean Sea. With a total length of up to 2 metres it is a relatively large stingray. In the North Sea it has been a rare summer guest, which seems to have changed in recent times. In June 2020, for example, a specimen of approx. 50 centimetres in length was caught by a shrimp trawler in the southern North Sea and documented photographically. Stingrays eat invertebrates and small fish. On his tail the stingray has one or two spines with a sawn edge with barbs. If he becomes attacked, he flaps its tail upwards and spears the attacker with his poisonous spine. However, he does not use this weapon for prey acquisition, but only defensively. Like other subtropical cartilaginous fish, he also benefits from the warming up of the Wadden Sea, because thanks to this warming, he may receive here a lot of food, especially true shrimps and small fish.

The sting of the stingray has many small barbs. In the middle you can see the poison gutter, which is connected to a poison gland. It is better not to get acquainted with it... The consequences are nausea, palpitations, diarrhoea and in the worst-case heart attacks and death.

The **Ray-Finned Fishes** include the vast majority of fish species that exist on our planet, and their physique is quite different from that of cartilaginous fish. Bony fish have an internal skeleton, that is connected in almost all parts. Their teeth are usually firmly anchored in their jaws, and their fins are supported by bony spines. In addition, most species of fish have ganoid scales all over their body, which are connected with a protective layer of mucus. There are only very few fish species without scales, including some catfish and some eel species. It is remarkable that there are also several species of fish, which have a real bone armour on their body, so that it almost seems as if they have an outer skeleton similar to that of some invertebrates. In the North Sea, these include species such as hooknose, seahorse or pipefish. In contrast to sharks and rays, almost all bony fish have a swim bladder with which they can regulate their buoyancy in the water. This is an organ filled with gas or oil, which can expand or contract according to the deep pressure of the surrounding water. Saltwater fish must constantly drink water because the salt contained in seawater constantly draws liquid from the fish's body. The excess salt in seawater is then excreted by the fish through special glands on the gills and kidneys. Freshwater fish, on the other hand, have to constantly excrete the water entering their body through their kidneys. That is why they urinate into the water. Some species of fish, especially migratory species such as sea trout or European eel, have no problem converting from seawater to freshwater and vice versa. While other species can only tolerate conversion to the other environment for a very short time. Some marine fish migrate to fresh waters to get rid of annoying parasites that die by changing the salinity, as their organism cannot tolerate the rapid change to other osmotic conditions. There are also freshwater fish that do the exact opposite. The perch, for example, is often found near the coast and sometimes even in fish traps in the waddens. The nine-spined stickleback can also be found in the waddens in rare cases, although it is actually a pure freshwater fish. With the fishes of the North Sea you can always have surprises, mainly because the fishes have not read the books written about them. One should never believe sweeping statements that have been made at some point by any "smart" people, and beware of generalisations. Also, in view of the debate on progressive climate change, it has become very difficult to assess, what the composition of the North Sea fauna will be in the future. For many species it is not yet clear whether their decline is "only" due to the warmer climate or simply to overfishing or pollution of their habitats. However, for some species it is quite predictable that warmer temperatures will force them to spawn in increasingly northern climates, as they cannot reproduce at higher temperatures. This is due to the fact, that many species require a cold winter period to mature their gonads (sexual products). If the cold phase does not occur, they stop reproducing. This problem is also known from commercial aquacultures and aquarium keeping of fish, but it does not only apply to fish, but also to invertebrates. At the same time, there are already clear trends, that more and more fish species, which actually swim in more southern climes, develop the north as a habitat. It should therefore be carefully observed, if species such as the striped red mullet, which is actually typically found in the English Channel and spawns in the southern North Sea, are suddenly found in the Elbe-estuaries. The reason, why such faunal shifts are so problematic is, that it is not possible to say exactly what effects they will have on the entire ecological structure of the North Sea. For the fishing industry, these changes can sometimes lead to the ruin of entire fishing fleets, but they can also create new opportunities. During the 1980s, for example, herring fleets were wiped out because of dwindling herring stocks, while fishermen, who switched to the sudden increase in mackerel shoals, benefited from this change. However, the situation becomes truly dramatic when, as happened in 2009, cod and flatfish stocks collapse for "no" apparent reason. Whether there will still be any profiteers seems questionable. And whether the immigrants from the south will be marketable also seems doubtful. Because customers usually consume, what they know. And not what they`re offered.

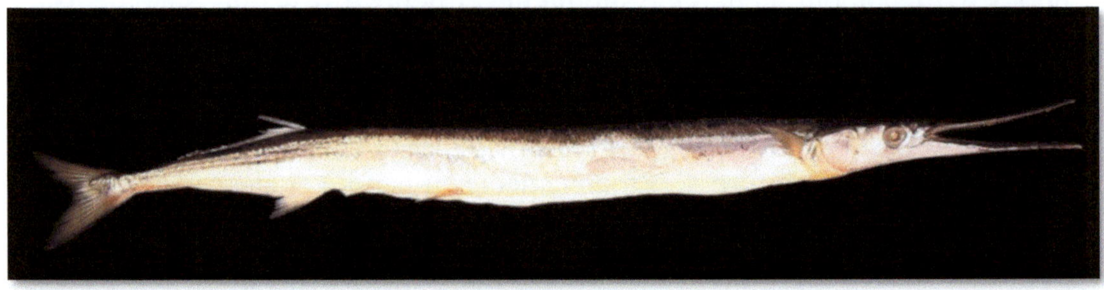

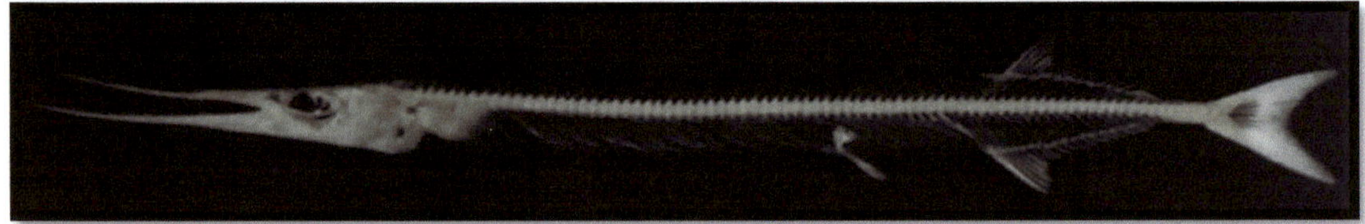

The **Garfish** grows to a maximum length of 90 centimetres and can reach a weight of 1.3 kilograms. It occurs in the Eastern Atlantic as well as on the coasts of the Mediterranean and Black Sea. The jaws of this fish resemble to the beak of a stork or heron. This "beak" consists of jaw bones, the upper part of which protrudes above the lower part and closes like a hinge. Garfish eat small herring and anchovies, which they chase in shoals just below the sea surface. The prey fish are first grabbed across the middle of the body and held in place with their pointed teeth. Then the upper and lower jaws are turned slightly apart horizontally so that the prey fish can be swallowed head first. Garfish are among the fish species, that will benefit from a slight warming up of the southern North Sea. The main reason for this is, that as the water temperature rises, their prey-fish will automatically reproduce faster. This is particularly beneficial to the growth of their own broods. It also has a positive effect on species such as garfish, when otherwise existing predators such as saithe and cod have to move northwards as a result of ocean warming. In this way, more specimens of their broods survive and quickly become adults. An increase in garfish stocks can therefore be expected in the foreseeable future. Not only in the North Sea, but also in the southern Baltic Sea.

The **Tompot Blenny** occurs regularly in the Mediterranean and Black Sea and the Eastern Atlantic up to the English Channel. And now he can also be found in the southern North Sea. And there are also unconfirmed reports from the island Heligoland. The species can reach a maximum length of up to 30 centimetres. These fish are found in association with seagrass-meadows or rocky coasts, where the juveniles can already be found in the shallow tidal zone, while adult specimens can reach depths of up to 30 metres. They feed on algae and the small animals living on them. The males occupy burrows where they spawn with several females one after the other to later guard the brood. After a few weeks, the brood swims in the plankton until the fully transformed young fish switch to bottom life. In 2016, a first juvenile specimen was found from the floating docks in the harbour of Baltrum. And in the following years the East Frisian shrimp trawlers caught single specimens. In spring 2020, one shrimp trawler even caught 6 specimens of this species. What proves, that this fish species is now no longer just seasonally present in the German Bight, but is already reproducing here and has thus established itself permanently. Based on aquarium observations, it can be said, that tompot blennies tolerate temperatures of 20° Celsius and more. They`re the true indicators and winners of the climate change.

The **Twaite Shad** is an anadromous migratory fish, which can reach a maximum length of about 60 centimetres and a total weight of up to 1500 grams. She occurs from Norway to the Mediterranean and Black Sea. Unfortunately, this species has become relatively rare due to construction and regulation of the spawning grounds. The twaite shad has one large and several small black spots on the lateral line, which are clearly visible only hours after the death of the animals. Freshly dead specimens glow a faint purple. Twaite shads reach with an age of up to 25 years a quite high age for a herring fish. For spawning they migrate into the river mouths, which was made impossible for them in many places with locks and weirs. In recent years, however, the cultivation of fish ladders and similar devices has begun to make it possible for migratory fish to reintroduce. However, progressive climate change could very quickly render all these human efforts ad absurdum, as herring like fish in Europe in particular depend on sufficiently large, oxygen-rich and easily accessible river mouths in which they can pursue their spawning business during May and June. If the climate becomes warmer, their spawning waters may no longer carry enough fresh water when water levels fall, so that they cannot proceed to reproduction. At present, catches of this once common herring species appear to become rare. And it would not come as a surprise, if these fish disappear completely from the German Bight in future. Whether it will also disappear from the Baltic Sea is unclear. But even here it has already become rather rare today (2020).

The **Herring** is a pelagic school fish that is always on the run. It grows to a maximum length of about 40 centimetres and can live up to 25 years. Herrings hunt bathypelagic copepods, which they follow on their day-night migrations from the surface to deeper regions. When they rise from the depths again, the herrings have to exhale small gas bubbles from their swim bladder, so that the difference in pressure does not tear them apart. These bubbles cause a bubbling at the surface, where the fishermen used to put out the shoals of herring. Herrings are excellent edible fish. The main advantage of them is, that they can be preserved in many different ways. The overfishing of herring stocks has led to a greater proliferation of herring's food competitors, such as the **Sardine *Sardina pilchardus*** and **Sprat *Sprattus sprattus***. The juveniles of the herring can also be found near the coast and in harbours during the summer months from around May. During the high summer it can happen, that a whole flock of adult herring strays into a harbour. Herrings require temperature ranges between 1° and 18° Celsius for their life and reproduction cycles. If the temperature in the southern North Sea rises too much, this is likely to lead to the complete disappearance of this commercially important fish species from this marine area in the medium to long term. It is highly unlikely that the herring will become accustomed to permanently high temperatures in the short term.

The **Sprat** is the smallest of the herring like fish species in the North Sea. It only reaches a length of about ten centimetres. The sprat occurs in a very large distribution area from the Norwegian Lofoten islands along the western British coasts. And from there to Morocco, the Adriatic Sea and even in the Black Sea. Sprats are caught frequently by shrimp trawlers during the summer months, but is usually discarded. But they are excellent as fish food or fishing bait. In the Baltic Sea they are also caught commercially and processed into smoked fish. Like all herring fish, they are very sensitive to the loss of individual scales and can therefore unfortunately not be landed alive for aquaristic purposes. Large sprats and small herrings look similar at first glance, but sprats have sharply protruding scales on the keel of their abdomen, which can be carefully felt and then seen on dead specimens. Also their body is a bit higher backed and stocky. Due to the very large distribution area of the sprat, it is likely to benefit greatly from a warming up of the southern North Sea, as its biggest food competitor, namely the herring, will most likely have to leave the field sooner. And since sprats can be counted among the Mediterranean fish species, they are unlikely to have any further problems adapting to higher temperatures. The sprat will therefore be one of the winners of the global climate change.

The **Sardine** is a common schoal fish found in the Mediterranean, the Eastern Atlantic and the southern North Sea. Sardines reach a length of about 25 centimetres and can easily be distinguished from the other herring fish of their range. This is, because their body structure is very stocky and thick when fed well. Furthermore they have prominent small black dots on their sides and also have a characteristic star-shaped pattern on their large gill covers. They also have relatively large scales, with about thirty scales in the lateral line. Sardines prefer warmer water, which is why they are only found in the North Sea at higher temperatures. In spring they migrate from the English Channel towards the Skagerrak, and in autumn they return south. This migratory behaviour is related to their own reproduction and the reproduction of their planktonic prey in spring. As a result of the warming of the North Sea, it is to be expected, that sardines will be caught here more frequently in future. It is even conceivable that, together with the sprats, they could completely displace the last herring from the southern North Sea.

The **European Anchovy** reaches a length of 20 centimetres, but it usually remains somewhat smaller. Similar to sardines, anchovies belong to the thermophilic species, which have their main distribution centre in the Mediterranean Sea and in the Eastern Atlantic south of the English Channel. At appropriate water temperatures and with a good planktonic food supply, they migrate northwards from April onwards. In exceptional cases, they can also reach the Skagerrak, the Danish Belt Sea and Scottish waters. Less than twenty years ago, they were so common in the southern North Sea in spring, that some ports had even set up for industrial processing. Now, however, they are only landed exceptionally, and the quantities landed do not justify industrial processing anymore. Anchovies have tasty but very salty flesh, which makes them unsuitable as fish food. Fresh specimens have a steel-blue shiny scale dress. In Europe there are several local tribes of anchovies, and worldwide there are said to be about 140 different species. The specimen shown here was caught by a shrimp trawler off the island Juist in April 2014. In total, several trawlers landed a several hundred anchovies here this season. The situation was similar in spring 2015, as the previous winter was very warm. At the same time, the trawlers also caught the **Sand-Smelt *Atherina presbyter***, which also started in April to move northwards from the English Channel. In retrospect, it must be said, that these animals migrated too early towards the north, which clearly supports the thesis of a too rapid global warming up. Such unpredictable migrations of commercially exploitable fish species provide us with an indication, that chaotic and anarchic conditions already prevail in the fauna of the southern North Sea today (2020).

The **Sand-Smelt** is a common schoal fish that lives near the coast and reaches a total length of about 20 centimetres. These fish are also known from Mediterranean and North African waters. That is why they have been classified as a subtropical species. However, sand-smelts prefer standing diagonally in the open water to lurk for small crabs. The main population of this species spawns in the English Channel, where they attach their two-millimetre eggs to seaweed and algae. Later, the spurges migrate further north, where they can even reach the Kattegat and the Danish Belt Sea. Sometimes they also enter estuaries and harbours to hunt for small crabs and the larvae of other fish. In the Mediterranean Sea they can often be found in shallow water. In the southern North Sea they are occasionally landed from March onwards, depending on the course of the previous winter, as a by-catch in crab fisheries, but they are economically insignificant. In late autumn 2019, some fishermen landed specimens of about 10 centimetres in length. That indicates, that they`ve now started staying in the German Bight for an increasingly long time.

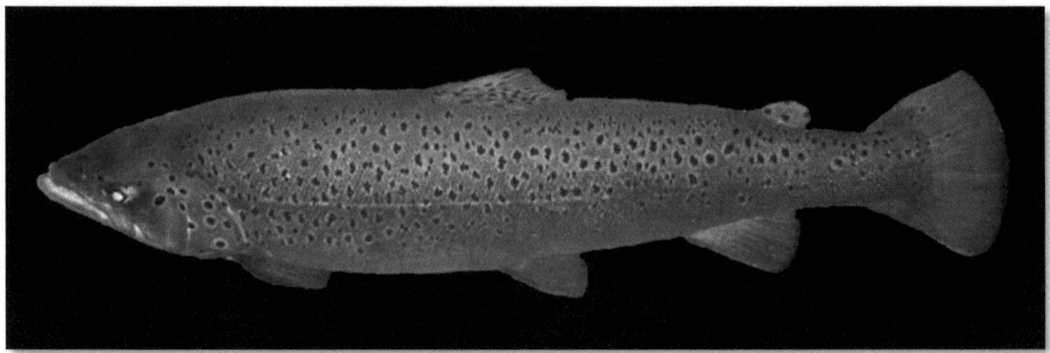

The **Sea Trout** is the saltwater variant of the brown trout. At its maximum, sea trouts can reach a length of 140 centimetres and a weight of 10-15 kilos. Sea trouts can be found from the Baltic to the Mediterranean and Black Sea in freshwater, seawater and migratory stocks. There are also populations in rivers, that do not migrate to the sea. The sea trout is an anadromous migratory form that lives in the sea and spawns in the gravel beds of rivers like the **Salmon *Salmo salar***. Overfishing is not a problem for such animals, as they are able to compensate this by producing vast numbers of juveniles. The main threat to the stocks of anadromous fish is obstruction of the spawning grounds, and pollution and destruction of their spawning areas. Pesticide inputs from agriculture can also have fatal consequences. Fortunately, trouts are already farmed in large numbers in aquacultures, so that the species as such is not directly threatened. Nevertheless, every effort should be made to preserve the wild stocks of this species. A further problem is, that trouts have also been moved to other parts of the world in the past, in order to make them usable for humans. This has led to massive faunal distortions and ecological damages in the affected areas, as trouts as active predators have considerably impaired the endemic fauna. The sea trout does not appear to be threatened by climate change for the time being, as it can also be found in the Mediterranean and the Black Sea. However, local tribes of this species could disappear and be replaced by others largely unnoticed.

The **Atlantic Salmon** can reach a length of up to one meter fifty and a weight of about 45 kilos. It was originally widespread and regularly found in all major rivers in northern Europe. Salmons are anadromous migratory fish, which migrate from the sea to the rivers to spawn and to provide for offspring in their smaller side arms and streams. Even in the 19th century salmons were so common, that employment contracts of Cologne employees stated, that they did not have to eat salmon more than five times a week. Due to industrial water pollution and the obstruction of spawning grounds, the salmon was almost completely exterminated, but the preservation of the species in aquacultures was successful. In the meantime, thanks to improved water quality, hatchlings have been released in the large rivers Rhine and Elbe, and initial successes in the form of adult salmons, which have risen to spawn, have already been recorded. For this purpose, special fish ladders had to be installed in many places to enable the salmons to ascend to their spawning waters. However, it will be many years before populations are established that can sustain themselves without human intervention. All these sincerely meant efforts will fall victim to the climate change, however, because subarctic fish species such as salmons can certainly neither tolerate nor adapt to summer temperatures of up to 28° Celsius (as measured in the Rhine in the heat year 2018). After all, salmons also belong to the subarctic faunal circle and will soon have completely disappeared from our latitudes under such increasing worse temperature conditions...

The **European Smelt** is a small coastal fish, that can reach a length of 35 centimetres. It is widely distributed along the North and Baltic Sea, and as a result of the last ice age there are also some pure freshwater stocks in various lakes. Freshly caught smelts smell of fresh cucumber, which makes these fish unmistakable. Smelts can live for about 6 years and become sexually mature in the sea at about 3-4 years, in freshwater at 1-2 years. Smelts are frequent anadromous schoal fish that migrate to the river mouths to spawn from March to May, where they spawn on sandy and gravelly surfaces. They have not or hardly been affected by human water regulation measures like other migratory fish, because they spawn in estuaries and not in small tributary rivers. Almost the entire abdominal cavity of the female smelts is then full of eggs. A female smelt can lay up to 50,000 yellowish eggs of about 0.6 - 0.9mm diameter, from which the brood hatches after 3-5 weeks. Smelts are fished and marketed locally as a speciality. After a harsh winter, their spawning migration may be delayed by a few weeks, leaving the fishermen empty. Nowadays the smelts have recently become rare in some estuaries, because too high water-temperatures have resulted in insufficient oxygen saturation of the water. In some places this phenomenon seems to have occurred in the context of water regulation measures. Often several reasons may cause the disappearance of such once common species.

The **Whiting** or **Merlan** is a small relative of the cod, which can reach about 70 centimetres in length. The whiting is widespread along the coasts of Northern Europe, the Adriatic, the Aegean and the Black Sea. On the German coasts whitings are usually caught by shrimp trawlers when they haul in the nets, as whitings often stay near the surface. Locally they are considered a speciality and are also prepared by the corresponding restaurants according to local recipes. They have good meat and belong to the best edible fish in the North Atlantic area. Especially in France they are marketed as merlan, which may lead to some confusion regarding their German name, but both names whiting and merlan are correct. Juveniles protect themselves from predators by staying between the nettle tentacles of jellyfish. Obviously, they possess, similar to the tropical anemone fish, a natural nettle protection in their skin, which protects them from the poison of the medusa. Whitings feed on small crabs and fish brood, which they catch near the coast. But they also eat the plant-like colonies of hydroid polyps, as I found out when examining the stomach contents of dead specimens. It is very likely, that whitings, due to their natural ranges, will benefit from the climate change. After all, if the cods have moved more and more northwards, they will have one less predator and one less competitor for food.

The **Atlantic Cod** is a predatory fish up to 1.50 metres long and can weigh unto 40 kilograms. It is a very important food fish, which can be found between 5 and 600 metres deep. It can be found both near the bottom and swimming freely in the open sea. It is a predator, that preys on other fish, crabs and mussels. The cod is not a true school fish, but he can sometimes be found in large numbers, if there is enough food. There are several different populations that live and spawn in certain areas. So the cods of the Baltic Sea do not mix up with those of the North Sea and are usually 20 % smaller. Since cod stocks have been heavily overfished and cods are now rarely taken by shrimp trawlers as by-catch (which once used to be somewhat different), it is possible, that the cod will one day be on the Red List of Threatened Species. A further problem for the cods is, that the progressive global warming up is pushing them further and further northwards, because they are one of the Arctic fish species that prefer temperatures between 2° and 10° Celsius. As a result, fishing trawlers are forced to travel further and further northwards to catch some cods at all. In the southern North Sea, the species has only been caught sporadically since the 2010s. And during the summer months it seems to be mostly absent in this sea area. If the cod is gone, it may be, that all human efforts to protect the climate have finally failed.

Saithe, *Pollachius virens* (Linnaeus, 1758)

The **Saithe** can reach a length of up to 130 centimetres and is usually found at depths of up to 250 metres. It becomes sexually mature at the age of 5 to 10 years. When the saithes have reached their spawning grounds north of Great Britain and off the Norwegian coast, they spawn at water temperatures of 6° to 8° Celsius at depths of 200 metres. The juveniles then live for up to 3 years near the coast, where they feed mainly on small crustaceans and fish broods. Then they migrate to greater depths themselves. The saithe is an arctic fish, which is mainly found and caught in the northern North Sea. The saithe is one of the economically most important edible fish in the North Sea, because he has a high-quality meat. Saithes are not yet as badly affected by overfishing as other fish species, as they have very high reproduction rates to compensate for their stock losses. But saithes have already become a rare exception in the southern North Sea and in the foreseeable future they will no longer be found here at all. Rare catches of this species, such as that of a single specimen by a shrimp trawler in November 2019 off the island Juist, paint a very schizophrenic picture of the North Sea`s fauna. This indicates, that in this small part of the world's ocean nothing is, as it should be...

With a final size of up to 45 centimetres, the **Pouting** or **Bib** is one of the smaller cod species. He is only used as an edible fish in Mediterranean countries, whereas the northern Europeans tend to process it into fishmeal. He is found in the North Sea, around the British coasts, in the western Mediterranean Sea. He is a rarer cod species in the North Sea. Poutings also live bottom-oriented and feed mainly on shrimps, small crabs, small squids and small fish. Juveniles often stay near the shore above sandy bottoms, while adults stay at greater depths unto 100 metres. They prefer protected places such as rocky reefs and shipwrecks. Poutings already reach sexual maturity at the age of one to two years and spawn in winter southwest of the British Isles and in the Mediterranean Sea. In the past, poutings were caught in the southern North Sea as a regular by-catch, but nowadays they are rarely landed. In addition, until the early 2000s, the juveniles could be caught during the summer on some East Frisian islands (especially on Borkum) with sinker nets near groynes or quay walls. In the meantime they have also become absent here. A consequence of climate change?

The **Burbot** is the only known species of the codfish-family, which can also live and reproduce in pure fresh water. Burbots occur in circumarctic areas, especially in the mountainous regions of Europe, North America and Asia. The burbot can grow to a length of about 150 centimetres, but is usually more like 40 centimetres long. It is found mainly in fresh water, but also in the brackish water of river mouths. In Europe they can be found in the mouth of the Oder, the Rhone or in the Loire. Especially during the winter period they can be found in estuaries because of their spawning migrations. Here they can be caught by anglers. Due to its need for cold water temperatures, this species is rarely seen in summer, but in winter or at greater depths. This is because burbots can also retreat to depths of 700 metres, making it very difficult to monitor or detect their populations. Locally, burbot populations are declining in many places, which could be due to pollution, climate change and overfishing. For this reason, burbots are also locally bred and released in tadpole-poor waters. With advancing global warming in the northern hemisphere of the earth, it is to be expected, that only small relic populations of this once widespread and common species will remain in extremely remote habitats. Such as in the Urals, Siberia or Alaska.

The **Five-Bearded Rockling** has five barbels, one of which sits on the lower jaw, the others on the upper jaw. With these tactile organs they track down small prey. This fish reaches a body length of about 25 centimetres and is therefore economically insignificant. It lives in shallow waters between algae stocks down to a depth of about 20 metres and is a regular by-catch in crab fisheries. Its colour is bronze, its scales are very small and therefore it makes an overall eel-like impression, which is further enhanced by its meandering swimming pattern. Adults spawn in the middle of winter, where they spawn at depths well below the tidal mark. They are fast swimmers, and the best way to catch some juveniles is, to use a frame net, which can be pulled quickly through an algae field in summer. They are usually found in stocks of sea salad. In shallow areas of the waddens you can find juvenile rocklings during the summer, that are usually 5-6 centimetres long and somewhat reminiscent of tadpoles. Their young animals even tolerate higher temperatures much better than adult specimens. Because based on aquarium and field observations I can say, that adult rocklings die within hours or days in 20° Celsius warm sea water. Therefore, even these actually common fish could become rare or die out in the future. Currently (2018-2020) I could observe a slight decrease of this species in the by-catch.

The **Four-Bearded Rockling** can reach a length of about 40 centimetres and is widespread in the North Atlantic, as it can also be found on the western side of the Atlantic from Newfoundland via Greenland and Iceland to Norway, Great Britain and in the entire North and Baltic Sea. It belongs more to the arctic species and prefers colder water. As its name suggests, it has a total of four beard threads, three of which sit on the upper jaw and one on the lower jaw. It can be found from depths of 20 metres down to 650 metres, where it hunts crabs and fish on soft or muddy substrates. This rockling is sometimes often landed as by-catch, but it is not used as an edible fish. The four-bearded rockling is good and durable in an aquarium, but it is one of the nocturnal and light-shy fish. That is why it is difficult to get to see it at all. Since about 2016 I noticed this species a little bit more often in the by-catch of the shrimp trawlers, while the five-bearded rocklings were caught less often. Whether climate change is responsible for this trend is unfortunately still unclear at the moment.

The **European Hake** is an interesting species, which can be found from the Norwegian fjords in the north to North African waters, in the Mediterranean Sea and occasionally in the Black Sea. In North Africa this important food fish is also known as **"Merluzza"** and is also offered and marketed under this name. The European hake is a very popular angling and food fish, which can reach a size of up to 140 centimetres. It is found at depths between 30 and 1000 metres, which suggests, that it is one of the bathypelagic fish species that follow the shoals of plankton and fish, on which they feed in the water column of the open ocean. This species is also fished commercially and is popular among sea anglers, which is why it is regularly offered for consumption. On the coasts of North Africa, a related species occurs, namely the **Senegalese Hake** *Merluccius senegalensis*, which however has an almost black back. This species is also frequently found off the coast of Morocco. If catches of hakes were to increase in our waters in the next few years, this would be a further evidence of the fact, that the North African fauna is shifting northwards. Similar to some swimming crabs such as **Pennant's Swimming Crab** *Portumnus latipes*, for example.

Ocean Sunfish, *Mola mola* (Linnaeus, 1758)

The **Ocean Sunfish** is a real cosmopolitan of temperate and tropical seas, which now and then also appears in the North Sea and Baltic Sea. It is a species of records, because it can reach a length of up to 3 meters and a weight of up to 1500 kilograms. In addition, it is probably the most productive fish species on our planet, as a female can spawn up to three hundred million eggs. Ocean sunfishes can be distinguished from all other fish by their characteristic caudal fin, which is more like a vertical fin seam. This makes the ocean sunfish look like a fish cut in half. The larvae of the ocean sunfish still have a normal caudal fin and several long spines as protection against predators, but both the caudal fin and the spines are transformed or receded as they grow up. Ocean Sunfishes feed on medusae, eel larvae and the larvae of squids, which they hold on to with their parrot beak and then incorporate. Ocean Sunfish are known to drift slowly on the surface, often letting their dorsal fins stick out of the water and sailing slowly. On this occasion, ocean sunfish have often been rammed by boats and have sometimes been captured. Unfortunately, ocean sunfish can only be seen alive in a few large aquaria in Europe, such as in Lisbon, for example. Since the 2000s, however, stranded ocean sunfish have been in the headlines again and again, with some specimens even reaching the Baltic Sea(!). The accumulation of such reports can be seen as an indication that the sea temperature has become warm enough for the stays of these subtropical animals, which now follow their food (jellyfish) further and further northwards into the world's oceans.

The **Thicklip Grey Mullet** can grow up to 75 centimetres. It is distributed from the Mediterranean Sea to the North Atlantic near Iceland, and can even be found in the south along the Atlantic coasts of Africa up to the equator. In the North and Baltic Sea, mullets can often be found, especially in the summer. Mullets are almost the only vegetarian fish living on the European coasts, feeding on algae and small particles which they filter with the thorns in their gills of soft substrates. Mullets spawn in the English Channel and near Ireland, as well as in the Mediterranean Sea. Their juveniles then migrate northwards and increasingly develop the coastal habitats there. Meanwhile, the broods of mullets linger longer and longer on our coasts. In November 2019 I was able to detect several 4-centimetre juveniles in the shallow waddens of Neßmersiel, on the coast of Lower Saxony. This is a clear proof, that the southern North Sea has already become much too warm in autumn and winter, because such catches should actually no longer be possible at this season in the shallow waters of the Wadden Sea. And it clearly proves that subtropical fish are now staying longer and longer.

European Seabass, *Dicentrarchus labrax* (Linnaeus, 1758)

The **European Sea Bass** or **"Loup de Mer"** can reach a length of one metre and a weight of up to 12 kilos. It occurs from Iceland in the north, around the British Isles, in the North Sea, in the Mediterranean and Black Sea and in the Eastern Atlantic Ocean south to Morocco. It can also be found in brackish estuaries and fresh water. As typical inhabitants of river mouths and lagoons, the young sea bass hunt invertebrates and small fish in small schoals. As adults, they live solitary and migrate during the winter to deeper zones down to about 100 metres. The sea bass is kept in aquacultures as a valued food fish. It is also known that sea bass can even be kept in warm waste water ponds of power plants. Sea bass are predatory fish that will eat anything that fits in their mouth. The juveniles can even be seen as surface-oriented small swarms in the ports of the southern North Sea during the summer months. They can be kept well in an aquarium at room temperature, as they are not sensitive to higher temperatures similar to Mediterranean fish. In recent years, sea bass have been on the rise, especially in the southern North Sea, which is clearly related to the warming up of this sea area. It has been proven that the North Sea has warmed up by at least 2°Celsius on an annual average over the last one hundred years. Since about the 2000s, sea bass have therefore not only been caught by shrimp fishermen in all sizes, but they are now also targeted for angling on the East Frisian Islands. Specimens up to 70 centimetres long and more have already been caught, yet.

The **Threespine Stickleback** grows to a length of 11 centimetres. This universal fish can be found in fresh, brackish and sea water. You can find it in Northern Europe as well as in the northwestern Mediterranean and the Black Sea. Sticklebacks are very adaptable, and you can even slowly change their habit from fresh to sea water. The threespine stickleback can be found in shallow water, where it`s juveniles appear in large shoals. Adults are loners. It means a scandalon, that a fish as common as the threespine stickleback is listed as an endangered species on some red lists, because it can even be found in habitats that are too small or too dirty for other fishes. Unfortunately its stocks are declining, especially inlands. The stickleback tribes living in our country can only withstand short-term warmth spurts in the summer tidal flats. If they are permanently warmed up, their life span is dramatically shortened. Should our native morphs of this fish species "disappear" one warm day, there is still the possibility, that heat-tolerant morphs from the south will move in and replace them. A loss, that could then only be proven by molecular genetic studies.

The **Sea Stickleback** which has 14-18 spines in front of the dorsal fin, can reach lengths of up to 20 centimetres. This species occurs only in sea and brackish water. It is found in shallow water between algae and seaweed, where its juveniles form swarms. During the reproductive period, the males of this stickleback have a yellow throat and belly, while the females are brownish. In all sticklebacks, the males build nests of algae, into which they drive the females to lay their eggs. Then the male guards the eggs until the brood has hatched. The sea stickleback is one of the fish species of the temperate climate zone, that is sensitive to heat and heat waves. It is therefore very likely, that this species will retreat northwards in the future.

Short-Snouted Seahorse, *Hippocampus hippocampus* (Linnaeus, 1758)

Short-Snouted Seahorses can be found on many European coasts and especially in the Mediterranean Sea, but they lack on so many British coasts and are completely absent in Ireland. They can reach a total length of 15 centimetres. The Latin-Greek genus name **"*hippocampus*"** literally means **"horse caterpillar"**. In fact, seahorses look similar to the jumper in a chess game, and in fact their prehensile tail with its bony armour is reminiscent of the body segments of a caterpillar. The fact that they are so agile and pliable despite their ossified appearance is truly amazing. A special feature is, that the females "pale" shortly before mating, indicating their mating mood. Then they lay their eggs in the brood pouch of the males and let the brood hatch out by the male. After a few weeks hundreds of fully developed juveniles hatch out. For many years' seahorses were considered to be extinct fish in the North Sea, but they had a delicate comeback in the early 2000s, especially in the southern North Sea. Here, single specimens are occasionally caught by shrimp trawlers, some of which also exhibited reddish hues. Seahorses benefit in any case from a warming up of the southern North Sea, provided, that this also allows their prey, small shrimps, to reproduce abundantly.

The **Greater Pipefish** is a rather large representative of its family and can reach a total length of up to 50 centimetres. It is very widespread, as it is found both near the Faroe Islands, around Great Britain, off the coasts of Norway, in the North Sea, in the Mediterranean and the Black Sea, in the entire Eastern Atlantic to South Africa and in the Indian Ocean unto Zululand along South Africa. Thus the greater pipefish is a cosmopolitan. This species inhabits habitats with algae stocks, but is lesser productive, as it only lays up to 200 eggs. The male also takes over the breeding business and carries the brood. With good feeding, the greater pipefish can be kept permanently, and I can confirm a keeping period of up to 6 years based on personal messages. At the beginning of the 1980s there was already a trend that greater pipefish became caught less and less by shrimp trawlers as by-catch. However, more recently, i.e. since the 2010s, it seems, that they are now being caught as by-catch more often than before. On the one hand, this could be related to the increased temperatures in the southern North Sea, but it could also be a consequence of protective measures or missing enemies. In addition, shrimp fishermen are very reluctant to fish in seaweed stocks, so that increased catches of pipefish in recent times could also indicate "acts of desperation" on the part of fishermen. In any case, these cute and graceful creatures benefit from a global warming up and will therefore be able to move further northwards in the future. The premise for this is, that they will always find enough food in the form of various small crustaceans.

The **Lesser Pipefish** can reach a length of up to 20 centimetres and is found in North and Baltic Sea. This species is usually found at shallow depths from the waddens down to about 20 metres. The lesser pipefish is one of the most common members of its family and can often be found associated to stocks of algae. Due to its greenish-brownish colouring, it is usually not visible from above through the water surface, and it is more likely to be caught accidentally with a frame net. They can also be caught in harbour basins with the baited gillnet, as they are attracted by the smell of fresh prey. The species can be kept and bred in an aquarium if it is fed with live food. The photo shows a male, which can be recognized by the clearly widened belly. Here a brood pouch is formed in which the female lays the eggs, which are then fertilized and carried by the male until the fully developed young pipefish hatch. The specimen shown here was found in June and was about to release the live young. At hatching they have a body length of just about ten millimetres. This pipefish benefits from the warming up of the seawater, which is why it can be found in late autumn, often until November, in the shallow water of the German Wadden Sea.

The **Snake Pipefish** reaches a length of 60 centimetres in the female sex and about 40 centimetres in the male. It is widespread and can be found from Iceland and Norway to the Azores. It also occurs in North- and Baltic Sea, where it can also occur in brackish water. Snake pipefish have neither pectoral fins nor do they have a tail fin. They are typical inhabitants of the algae zone, and since fishermen do not like to collect algae stocks with their nets, they rarely end up in the by-catch. Snake pipefish typically sway back and forth between the algae in the flow, hoping that small crustaceans or fish brood will swim right in front of their mouths. These are then literally pipetted in by cleverly folding the hyoid bone due to the resulting negative pressure. Snake pipefish are productive animals that can lay up to 1,000 eggs per brood. The males also carry their brood in a two-part brood pouch, which they carry under their body. After a few weeks, the juveniles are born alive and released into the wild. While big snake pipefish were rather declining in the 1980s, they are now, similar to the greater pipefish, caught more frequently again. This could also be an indication of increased sea temperatures, as the heat can also enormously accelerate the development of some fish broods depending on the reproduction rates of the plankton.

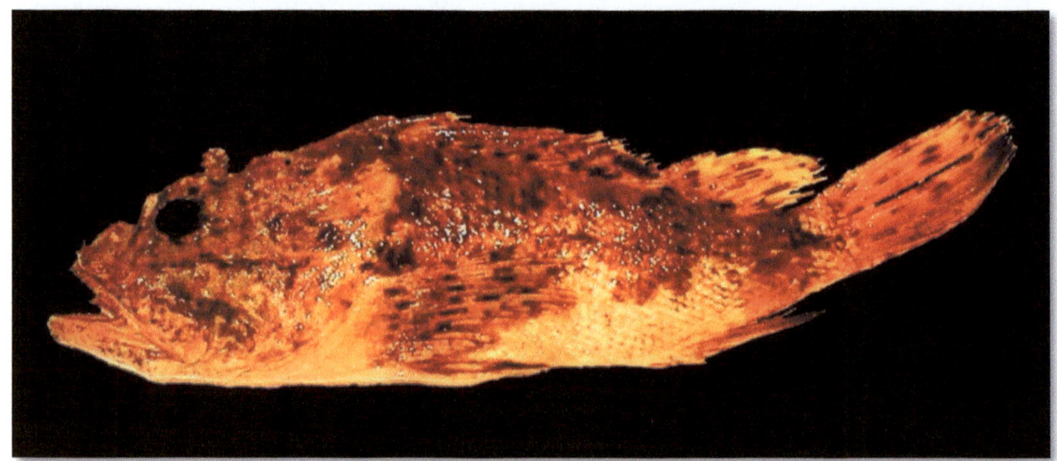

The **Red Scorpionfish** is a typical sea scorpion, that occurs in the Mediterranean Sea, on the North African coasts, on the south-western coasts of the British Isles and more rarely in the English Channel. These fish prefer depths between 5 and 200 metres. The species can reach a size of up to 50 centimetres and is despite its toxicity a highly valued food fish in Mediterranean countries. And which is important in the traditional *bouillabaisse*. The red scorpionfish can be distinguished from other species by the skin appendages under the lower jaw and the small brush-like tuft of skin above the eye. This species has poisonous fin and gill-cover spines. It should be noted, that it is considered the most poisonous species of all European scorpionfish, and that its venom is still effective after the animal has died. Children and allergy sufferers are at risk of death after a sting, which is why a respectful handling is always necessary. The red scorpionfish lives associated to rocky substrates or reefs, often occupying exposed viewing spots. Here he lurks for small careless fish and shrimps. With an increasing warming up of the North Sea, it is to be expected, that species such as the red scorpionfish could establish themselves permanently on the rubbish reefs in the southern North Sea, and thus may become a regular by-catch animal in the German shrimp fishery.

Blackbelly Rosefish or Blumouth, *Helicolenus dactylopterus,* Delaroche, 1809

The **Blackbelly Rosefish** or **Blue Mouth** is an inhabitant of deeper areas and usually lives at depths between 50 and more than 1000 metres. Until yet, this species was known from Norway, the western British Isles and from African coasts. This perch-like scorpionfish can grow to a length of about 50 centimetres and belongs to the redfish family. And like the latter, the blue-mouth has poisonous fins and gill-cover spines. In November 2019, a single specimen of about ten centimetres in length occured in the by-catch of a shrimp trawler off the island Juist; in March 2020, several more specimens of the same size were caught nearby. It was unusual, that all catches were made in shallow water between 5 and 20 metres deep. And that this fish had not been known from this area before. Until yet, these fishes were known from the English Channel or from fjords of Norway. Thus, this species is an example for the fact, that the ichthyofauna of the English Channel, including its deep living species, shifts northwards. This is probably due to the fact that fish mainly follow the migrations of their prey, such as in this case the sand-smelts, which also occured in the by-catch during this period. And these followed the migrations of small crustaceans in the water column, because the warming up had favored their rapid reproduction.

The **Red Gurnard** reaches a final size of up to 70 centimetres and can be found at depths of about 15 to 400 metres. At first sight it can easily be confused with the **Tub Gurnard *Chelidonichthys lucerna*, but unlike this one, it never has blue colour patterns on its pectoral fins. It is also longer in build and not as stocky as the tub gurnard. Its colouring can be reddish or sometimes yellowish. The red gurnard is also offered in the food trade. It has good, but unfortunately bone-rich meat. Often, freshly caught gurnards also contain undigested prey from the sea bed, such as hermit crabs and swimming crabs. So far, this species has been detected from the Belgian coast, but it is conceivable that it will appear regularly in the German Bight in the foreseeable future and establish itself there with a permanent population. Gurnards are fish, that constantly follow their prey in little swarms. And as this is mostly made up of true shrimps, hermits, swimming crabs or small fish, they are often landed by shrimp trawlers as by-catch during the summer months. So in the foreseeable future, one can expect to catch more and more red gurnards in the German Bight.

The **Tub Gurnard** reaches a length of up to 75 centimetres and a weight of up to 6 kilos. It is widespread in the Northeast Atlantic and is found in the North Sea, the Mediterranean and Black Sea and along African coasts up to Mauritania's Cape Blanc. This gurnard lives bottom-oriented, chasing small fish, crabs and molluscs, but it can also be found free-swimming in open water, where it hunts mainly small fish such as sprats, sardines and similar. An anatomical peculiarity of this species are the three finger-like rays of the pectoral fin, with which it seems to "walk" over the seabed. With these it feels prey, and possibly tastes it with them. By the way, gurnards owe their name to the fact, that they can actually make growling noises with their swim bladder, if you take them out of the water. It's a fascinating experience if you've never experienced this before. Gurnards tolerate too warm water only for short terms of a few days. If the water gets warmer than 20° Celsius, they migrate into deep- and cold-water layers, as these contain more oxygen due to the density anomaly of seawater. This is the only way to explain why tub gurnards were almost never caught by shrimp trawlers in the shallow waters of the southern North Sea during the extremely hot summers of 2018 and 2019. And other species did the same to the tub gurnard, so that the shrimp fishermen caught almost no more fish as by catch in the summers of 2018 and 2019.

The **Grey Gurnard** bears this name wrongly, because living specimens shine bronze to gold and are anything but grey! It can reach a maximum length of 60 centimetres and a weight of about one kilogram. It is found from Iceland in the north to Morocco in the south. It can also be found in the North Sea, the Mediterranean and Black Sea. The grey gurnard lives in small groups on soft soils, more rarely on stony substrates, where it can reach depths of more than 300 metres. It has specialised in capturing small bottom fish such as flatfish, gobies and sand eels, but also small herrings. Shrimps and small crabs are not safe from him either. The grey gurnard is one of the fish species, that have a bathypelagic way of life, because at night they rise up to the surface where they followes their prey. At the beginning of the 2010s, these gurnards were occasionally caught by shrimp trawlers off the East Frisian Islands in spring at depths from 10 metres, and in summer they were not actually caught at all. In the meantime, i.e. now in spring and summer 2020, juveniles of about 10 centimetres in length have occasionally been caught in shallow water. This is an indication that there must have been a fundamental shift in the species composition of the southern North Sea.

The **Striped Red Mullet** reaches a length of up to 40 centimetres and can weigh up to one kilo. It belongs to the species of the Mediterranean faunal circle and usually occurs in the Mediterranean and Black Sea, on the Canary Islands and in the Eastern Atlantic up to Senegal. In the North Atlantic it occurs from the British Isles to Norway. Striped red mullets live bottom-oriented and can be found from shallow water to depths of 60 metres and more. They swim across the bottom in small groups and use their chin beards to feel out small prey from the bottom. You can find them on hard ground as well as on sandy or muddy ground. The striped red mullet should only rarely be found in the German Bight as a migrant, but due to the global warming up it has become a constant guest in the German Bight and can now frequently be found in places such as the Elbe estuary. Since the 2010s, it has also been regularly caught by German shrimp trawlers from May to December. Nevertheless, the striped red mullet is sensitive to high temperatures, although it belongs to the Mediterranean faunal circle. If it gets warmer than 20° Celsius, it dives into deeper water layers like the gurnards.

The **Atlantic Horse Mackerel** can grow to 70 centimetres long and weigh about two kilos. Horse mackerels can be found along the entire East Atlantic coasts from South Africa unto the Mediterranean and the Black Sea. The horse mackerel is unmistakable because it has bony scales along the lateral line on the back half of its body. It also has a characteristic black spot on its gill cover. Horse mackerels live as schoal fish above sandy substrates, where they hunt small crabs, squids and small fish. Young horse mackerels live pelagically between the umbrellas of medusae, where they eat not only plankton but also the genitals and tentacles of jellyfish. Horse mackerels are used as food, especially in southern Europe. They can therefore also be obtained fresh or deep-frozen. They have good, but bone-rich meat. In recent times, catches of this species have not necessarily increased, which could be due to the food competition from other fish species that have migrated to the southern North Sea. Should they be caught more frequently in summer in the foreseeable future, this would not be surprising for such a subtropical fish species. And should they hibernate in the North Sea one warm day, then all climate protection measures can be regarded as failure…

The **Morocco Dentex** belongs to the tropical and subtropical faunal circle, which is otherwise known from the African coasts, the Mediterranean and the British coasts. It has rarely been found north of the English Channel yet, but has been caught sporadically in the Kattegat. In the foreseeable future, this species could appear more frequently in the southern North Sea, as the internet platform fishbase.org already indicates (as of July 2020) that this is currently still a remote possibility. It can therefore be assumed, that this species will first pass through the English Channel to the north and then, due to the lack of hard substrates, will be found associated to the offshore wind turbines at Borkum island. This is, because all sea breams are usually found near rocky or steep coasts. Unfortunately, fishing is not allowed near offshore installations, so it will be difficult to detect this species. The morocco dentex has long been known to enter the north on the back of the warm Gulf Stream. This is especially true, if warm periods last for particularly long periods. Presumably, they are sometimes simply forced to develop other habitats when they have reproduced too much and their original habitats have become too small for the entire population. Their future in the German Bight therefore seems very likely and only a matter of time.

The **Salema** occurs in the Mediterranean Sea, on the North African Atlantic coasts and has rarely been detected in the southern North Sea and even from Swedish waters in the Baltic Sea. This species can already be found in shallow water in large swarms, where they cavort in the light-flooded water. Here they mainly graze on algae, and they can even eat thick-leaved seaweed. They are very peaceful and sociable fish that also like to swim with other sea-breams. Especially in the Mediterranean area these fish are marketed as food from a size of about 10 centimetres. So far unconfirmed single catches of this species by shrimp fishermen from the year 2019 can be assumed as probable. It may be assumed, that this species will use the offshore wind turbines at the island Borkum as a springboard to the East Frisian Islands. And it should therefore come as no surprise to us, if it is already cavorting in front of the only rocky German island, Heligoland.

The **Gilthead Seabream** belongs to the sea bream family and can grow to 70 centimetres long and weigh up to 17 kilos. It is from Norway around the British Isles to the Strait of Gibraltar and the Canary Islands, and throughout the Mediterranean and Black Sea. In between it has also been sighted near Heligoland in the summer. Gilthead seabreams live at depths of up to 150 metres, but they usually stay at a depth of about 30 metres and in spring they enter estuaries and lagoons to spawn. They feed mainly on molluscs and even thick-skinned oysters. They crack open these with their strong jaws and then eat the soft innards. The animals can produce impressive cracking sounds in the process. The gilthead seabream is mainly bred in aquacultures in Greece and is a commercially important species, which is traded as **"Dorade royal"** or **"Dorade grise`"**. Almost all of the sea breams sold in the fish trade originate from these aquacultures, but in the near future the supply of wild caught specimens, which landed as by-catch in the nets of shrimp fishermen in the southern North Sea, could increase. Since the middle of the 2010s, I have repeatedly heard reports of single sea-breams, that have been caught by various fishermen. It can therefore be assumed, that the gilthead seabream is already present in the German Bight every summer. If this species were to reproduce one warm day in the German Bight, this would be a very clear indication of the enormous progress of climate change.

The **Lesser Weever** reaches a maximum size of about 15 centimetres. It already occurs in shallow water and can be dangerous for bathers. Its poison is stronger than that of the **Greater Weever *Trachinus draco***, which is why it should be taken seriously by fishermen, anglers and bathers. This is because the lesser weever has poisonous spines in the first rays of its dorsal fin and on the gill caps. In order to avoid poisonous accidents, anglers should be extremely cautious when checking off angled specimens, and it is better to wear bathing shoes when bathing in risk areas. Especially in the Mediterranean area this often makes sense just to avoid injuries caused by sea urchins on gravelly ground. The way of life of the lesser weever corresponds largely to that of the greater weever. Since about the summer of 2016, this fish has been landed relatively frequently as a by-catch by the shrimp fishermen off the East Frisian Islands. One can rather attribute the lesser weever to the subtropical faunal part, which now slowly pushes northwards into the German Bight. This is a worrying development, because such animals could cause serious incidents, if they are more frequent on a warm bathing day in the shallow waddens.

The **Greater Weever** belongs to the thermophilic fish species and can be found from Norway to Morocco, on the Canary Islands and Madeira and in the Mediterranean and Black Sea. It reaches a length of up to 45 centimetres and can actually only be described as a toxic species. This is because the family of the *Trachinidae* consists exclusively of members, whose first dorsal fin is equipped with poisonous spines, which are preferably erected when a bather steps on the fish buried in the sand. Although the poison is not lethal, it is very painful and leads to cramps, vomiting and circulatory problems. Such wounds must be rinsed out as hot as possible and treated medically, otherwise victims may get health problems for several weeks. As this fish can be found on sandy bottoms already from a depth of 1 metre water depth, bathers should rather wear bathing shoes to avoid accidents. But also fishermen and anglers should better have a holy respect for the spines. During the day, greater weevers dig themselves into the bottom gravel, because they can wait so comfortably until small fish and crabs swim right in front of their mouths. When burying themselves they make meandering movements and in a few seconds they`ve almost completely disappeared. In recent years, greater weevers have been caught more frequently in the North Sea, which deals with the warming up. These fish can therefore also be seen as indicators of climate change. Nowadays they are currently collected by fishermen's nets in German waters much less frequently than the lesser weevers. Possibly due to the competition between the two species.

The **Atlantic Bonito**, is a migratory fish found in the Atlantic, the southern North Sea, the Danish Belt Sea, the Mediterranean and Black Sea. It can reach a length of almost one metre and penetrates from the water surface to a depth of about 200 metres. This gives it a bathypelagic lifestyle, in which it follows the migration of its planktonic prey in the water column depending on the daytime. Bonitos are excellent food fish, which have similar meat to their larger relatives, the tunas. They are commercially caught and mainly processed into canned fish, which is then marketed as ***"tuna"***. This is done, because tunas have now become a rare resource that people would rather market elsewhere, and especially at very high prices, on the Japanese market as sushi. If you want to contribute to the protection of species yourself, you should therefore refrain from such exclusive delights. Canned tuna can be consumed with a clear conscience, however, because it is now obtained exclusively from bonitos and therefore no stocks of the already overfished genuine tunas are endangered. Since about the 2015s, there have been more and more reports from fishermen who claim to have caught ***"tunas"***. However, it can be strongly assumed, that these catches were in fact large bonitos, as there are certain similarities. The occurrence of such Mediterranean species is an indication of a rich food supply in the North Sea, which would probably not exist in this form without climate change.

The **Atlantic Mackerel** is a shoal fish up to 60 centimetres long and up to 3.4 kilograms heavy. Mackerels feed on plankton, which stays in the open water in summer. The mackerel occurs from the North Atlantic, the North and Baltic Sea unto the Mediterranean and Black Sea. Mackerels mainly eat small invertebrates and brood, which they filter out of the water with their gill traps. They themselves serve as food for other large predatory fish. An anatomical peculiarity of mackerels is, that they do not have a swim bladder. But they owe fat meat, so that they get some buoyancy in the water due to the fat parts of their tissue. Mackerels eat fat pads in summer, which can be seen particularly well in autumn. In winter, the mackerels move into deeper water layers where they stop feeding and expect the next spring. This is due to the fact, that they have adjusted their life rhythm exactly to the growth and reproduction of the animal plankton, and in winter they cannot get enough planktonic food from the surface. In 2014 I received informations from various sea anglers, that they were unable to catch mackerels at all this year off the East Frisian Islands. This is probably due to the fact, that the year 2014 was one of the warmest since the beginning of the measurements of water temperatures in the North Sea. The absence of mackerels is probably related to the migrations of their planktonic nutrients and the oxygen demand of these fish, which cannot be saturated if water temperatures are too high. Interestingly, in both June 2016 and June 2020, shrimp fishermen caught huge quantities of juvenile mackerels - after a temporary cold spell!

The **Goldsinny** is probably the most common wrasse in Europe. It can reach a length of 18 centimetres and live up to 8 years. It can be found from Norway to Morocco, in the North and Baltic Sea, and in the Mediterranean and Black Sea. It can be easily distinguished from other species by the two black spots at the beginning of the dorsal fin and the caudal fin. As the goldsinny prefers rocky substrates, it is rarely found in the Wadden Sea, but more often off Heligoland, on the Danish and British rocky coasts and in the Baltic Sea. Goldsinnys eat bryozoans, crustaceans and snails. They can be found down to a depth of about 50 metres. Goldsinnys are territorial fish that defend their territories against other specimens in order to enable their own breeding business. In summer, the males guard the brood and defend it against attackers. The juveniles hatch out of the algae nests after one to two weeks and then live planktonic, before they change to a bottom-oriented way of life like the adults. Due to the newly created artificial rock bases on the offshore wind turbines, catches of goldsinnys off the East Frisian Islands have increased slightly since the 2015s. In addition, they seem to tolerate temperature increases up to the mark of about 20° Celsius.

The **Lesser Sand Eel**, also known as **Raitt's Sandeel**, can grow up to 20 centimetres long. It lives mainly in the Northeast Atlantic, but is also present in some parts of the northern Mediterranean and especially on the Spanish coasts. In summer, it can also be found in shallow water from the tidal zone onwards. In winter it retreats to depths of 20 to 50 metres. This species is characterised by its hydrodynamic streamlined shape and the meandering way it swims. Despite its small size, the lesser sand eel is a commercially important species, which is landed in huge quantities, especially in Denmark. The animals are then processed into fishmeal or animal feed. The lesser sand eel is a typical inhabitant of the sandy bottom and can burrow into it at lightning speed when in danger. At low tide, the fish may even be left in the dry sandy bottom of the tidal flats or in small puddles. I myself accidentally kicked such a buried specimen to death a few years ago. The lesser sand eel belongs to the extremely oxygen-dependent species of the tidal zone and can only survive here for a very short time in small tidal pools, which have become too warm. It also usually does not survive for long transports. However, it is remarkable that in some areas, such as on the Schleswig-Holstein North Sea coast in the northern German Bight, it can be found for longer and longer periods near the coast. Thus, in the meantime, one can find this fish even still in October and November in tide pools at the beaches of St. Peter Ording, for example. I was able to see this personally, especially after the heat summer of 2018.

The **Black Goby** is widespread from the Baltic Sea to the Mediterranean Sea and can be found particularly often in shallow harbour basins with sandy or muddy bottoms. With a total length of up to 10 centimetres, it is one of the larger and predatory goby species and is therefore often the dominant species. It can live at least 4 years. Adult black gobies can often be found in pairs; the males differ from the females by having a slightly higher first dorsal fin. Black gobies prefer somewhat shaded hiding places, from which they can be lured out with appropriate baits. They are then relatively easy to catch with a bait trough or a frame landing net. From May to August, black gobies lay up to 6,000 eggs on stones or seaweed. The males inseminate and guard the brood until it hatches. The larvae hatch out of the pear-shaped eggs at a length of about 3 millimetres and then swim freely. Only at a length of about one centimetre they switch to bottom life. Black gobies need 2 years to become sexually mature. In the southern North Sea, this goby has been rather rare, but since the 2015s it has been caught occasionally by shrimp trawlers. It might happen, that the combination of artificial habitats at offshore wind farms and a warming up of seawater will help this goby to spread more successfully in the southern North Sea in the closer future.

This invasive species originally came to our waters from the Black Sea and has now reached several rivers such as Danube and Ems and the Baltic Sea. It has become widespread through ships and has now even been detected in North America. It is a very universal fish that can be found in fresh, brackish and sea water. It grows to about 20 centimetres. The **Round Goby** has become a major pest of our endemic fauna, as it displaces our native species. It achieves this on the one hand by high reproduction rates and on the other hand by eating the broods of other fish. As a result, species such as the **Black Goby *Gobius niger*** have already fallen far behind. However, cold winters could well limit the advance of such species from overseas. The development of their populations should therefore be carefully monitored by the fisheries authorities. Their unchecked spread can have dramatic consequences for endemic fish species and the local fishing industries based on them. In the aquarium, these gobies can be kept well and persistently at room temperature without any problems. Their strong adaptability is probably the secret of their great success. And their aggressive progression combined with the climate change could mean the end for our endemic fish.

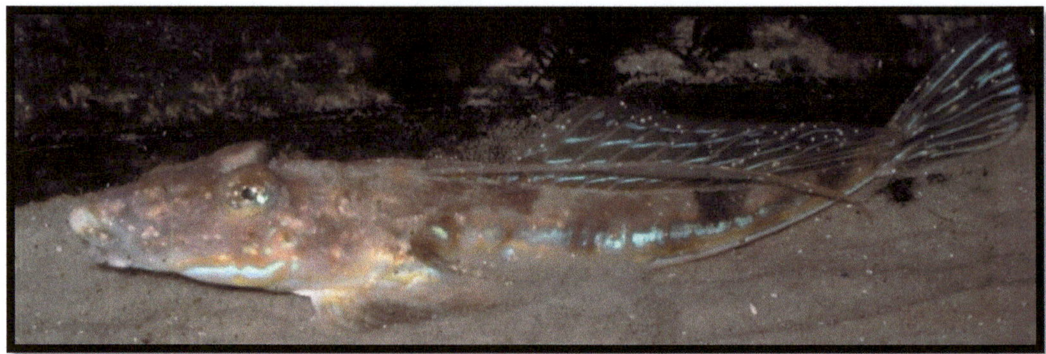

The **Dragonet** belongs to the first European fishes, that were kept in aquaria during the 19th century. These fish also occur in the Mediterranean Sea, where they usually live in deeper zones than in the North Sea. This species shows some astonishing adaptations to life on sandy substrates, which are comparable to those of rays and flatfish. The respiratory water behind the eyes is passed over the gills through a special sealable breathing hole, to prevent sand from entering the gills, when the fish buries itself. Dragonets also have a sting on the gill cover, which is connected to a poison gland and releases a weak poison. Therefore, you should not catch them with a net if possible, because they could easily get caught in it with their gill cover spines, what may cause enormous damage to the fish. Since dragonets live in the sublittoral below the tidal mark, they avoid the warm shallow and only tolerate temperatures up to 18° Celsius. During the warm summers of the years 2018 and 2019, no dragonets were caught by the shrimp trawlers, as the warming of the southern North Sea to temperatures of up to 22° Celsius had forced them to dive into deeper oxygen-rich waters. In early summer of 2020 they were sporadically present again, as it was obviously cool enough for them again. This could have been a direct consequence of the involuntary climate protection measures taken as a result of the corona epidemic, as not inconsiderable parts of global air traffic were suspended for months. It is truly amazing, what some strong measures could achieve in the short term.

Swordfish, *Xiphias gladius* (Linnaeus, 1758)

The **Swordfish** is found in subtropical waters in all oceans of the world and is heavily fished commercially in the Mediterranean. Nowadays it is also caught in North - or Baltic Sea. Or stranded there, especially near offshore islands. It is controversial among experts, whether the sword of the swordfish really only serves the purpose of prey acquisition, or if it just improves its hydrodynamics. But it is a fact, that swordfish are among the fastest fish of all, as they can move forwards 100 metres in just 3 seconds. They are dangerous for anglers as long as they are in the water. Because they can mobilize all their power reserves again shortly before landing and jump. Anglers are also at risk of being wounded into the fishing line and torn into the depths by the swordfish's extended upper jaw. Since swordfish can dive down to a depth of 800 metres, a fisherman, who is carried away, has no chance of survival simply because of the sudden pressure of the depth. Swordfish have already been severely overfished, and as they are at the end of the food chain, it is questionable whether eating their meat is a healthy consumption, as heavy metals such as mercury and cadmium are deposited in it. In pregnant women, the consumption of such mercury-containing food can even damage the foetus directly! More recently, swordfish found dead on the Baltic Sea Island Rügen (as happened in 2008) or in Wismar (2020) made headlines. There are also reports of other specimens that were stranded on the East Frisian Islands in the southern North Sea in the early 2000s. These specimens can be seen as further evidence of a progressive climate change, which should make us very thoughtful.

The **Plaice** is not only common in the North Sea, but also in the north-western Mediterranean Sea and on the coasts of North Africa. Although plaices can grow up to one metre long, such fish have become a rare sight due to overfishing. It should be noted that the flatfish stock in the southern North Sea has actually always been relatively good, despite intensive fishing, but it completely collapsed in 2009. This is most likely due to the **Rib Jellyfish *Mnemiopsis leydyi*,** introduced from the Black Sea, which are very fond of exterminating fish broods and can snatch whole fish stocks away. In the meantime (2020) the stock of plaices in the southern North Sea has declined sharply, which some fishermen see in connection with the power lines of offshore wind turbines. It is a fact, that juvenile flatfish are currently almost impossible to find in the waddens. And if, then only sporadically for a short time. It is unfortunately not known whether they react to excessive warming or other factors. However, their large-scale disappearance is a very strong statement from Mother Nature to the people.

Flounder, *Platichthys flesus* (Linnaeus, 1758)

The **Flounder** is a common flatfish, which can grow up to 60 centimetres long and weigh 2.5 kilograms. They are found in the North Sea, the Baltic Sea, the north-western Mediterranean and Black Sea. Up to now, flounders have been found most frequently in the area of the Baltic Sea, as this flatfish prefers lower salt levels. However, in spring 2020, flounders were caught much more frequently than usual in the southern North Sea, which may also be due to the decline of the plaices. In addition, flounders also often migrate temporarily into fresh water, so that they can even be fished inlands. This is known mainly from rivers like Elbe and Schlei. In the Baltic Sea, flounders are fished commercially and mean an important commercial fish for small local fisheries. Flounders have excellent meat and are one of the best edible fish in the North Sea and especially the Baltic Sea. The reproduction of its stock in the southern North Sea is an indicative of chaotic and unusual new underwater environmental conditions. Obviously, the entire ichthyofauna of the southern North Sea is undergoing a tremendous change.

The Solenette grows to a length of only 15 centimetres; it is found in the Mediterranean and North Sea. It is a regular by-catch.

The Sand Sole reaches a length of 40 centimetres and occurs in the English Channel, on North African coasts and in Mediterranean areas. In 2013 a shrimp trawler from Norddeich caught some juveniles…

The **Sole** is a seasonally common flatfish, which can also be seen swimming freely near the surface during its spawning season. This popular food fish is traded at high prices. Sometimes miscoloured soles are caught, and there are both albino and melanistic animals. The sole is characterised by the peculiar shape of its mouth and the diamond-shaped scales. They can feel their prey with the short fringe-like appendages on the hem of their head. Soles are specialized in eating mussels, worms, brittle stars and other small invertebrates. Soles can live up to 20 years, reach a body length of 60 centimetres and a weight of up to 3 kilograms. However, such large specimens have become rather rare due to general overfishing. In addition, it should be noted, that fishermen in the southern North Sea are currently not making the effort to actively fishing for soles, because the negative stock situation means, that the catch is hardly profitable at present. It remains to be seen whether this is the result of overfishing or climate change, and no clear answer can be given at present. It is also conceivable, that several species from the large sole family are currently competing for the habitat in the north, as they occupy the same ecological niches and have very similar nutritional requirements. Outcome uncertain.

The **John Dory** reaches a maximum length of 90 centimetres at a weight of about 8 kilos. It is a true cosmopolitan, which can be found in all temperate oceans. It can be found in the North Sea as well as in the Mediterranean Sea or offshore from Japan or Australia. The ends of the dorsal fin of the John Dory end in long threads, but the most bizarre feature of this fish is a series of oversized and spiky scales under the second dorsal fin. These bony plates are reminiscent of the bony side shields of sturgeons. Thus the John Dory enjoys a passive, but very effective protection against larger predators. This species is a loner, which is rarely found in small groups. He feeds mainly on small fish, but occasionally he also eats small crabs and squids. His hunting tactics are, that he stands completely still on the bottom and waits for a victim to swim right in front of his mouth. This then will suddenly be turned out, and the prey is ingested by the resulting negative pressure. As the John Dory has a very flat body profile, he is only visible to its prey from the front as a thin line, that resembles to a kelp-string swaying in the flow. In the southern North Sea, this fish has only been caught very rarely so far, such as one single specimen during the hot summer of 2016.

Anglerfish, *Lophius piscatorius* (Linnaeus, 1758)

The **Anglerfish** is a very bizarre looking fish, which can reach a maximum length of about 2 meters and a weight of up to 40 kilos. Anglerfishes do camouflage themselves by body appendages that look like small pieces of algae. This enables them to mimic algae stones, but also almost completely dissolve their body contours. Anglerfishes penetrate from the shallow coastal waters to a depth of 1,000 metres. Since they do not have a swim bladder, anglerfishes are rarely seen swimming freely. Between their eyes they have a freely moving dorsal fin, which it can swivel back and forth in front of its mouth like a worm. If a smaller fish is interested, the anglerfish opens his mouth in a flash and sucks in the prey. The victim has no chance, as the anglerfish's teeth are all directed backwards, what means no escape. In November 2019, a shrimp trawler for the first time caught several small anglerfishes of only 20 centimetres in length off the island Norderney in shallow water. Anglerfish of this size are normally only caught in the English Channel, whereas they are landed larger in the deeper North Sea.

I am not a scientist who can present meticulously determined facts using strictly scientific methods. But I would describe myself as a very communicative person, who above all has good powers of observation. And who is able to relate the most diverse things to each other. Since he has a very large background knowledge on the topic described, which has matured over decades. It's like putting together a huge puzzle from millions of different sized pieces. But of course you can only handle a tiny fraction of these pieces and try to construct tiny sections of a gigantic picture. And in our case concerning the marine fauna in the southern North Sea, this can only mean that an ambiguous distortion of what one would have to call a shift of the otherwise usual faunal parts in this small part of the one big ocean is created. In other words, the fauna of the English Channel and the British coasts is beginning to establish itself actively and permanently in our latitudes, the German Bight. In addition, finds of subtropical and tropical species are accumulating, which are entering the German Bight as a result of warm currents, lack of winter and other favourable conditions. Should a great white shark enter the mouth of the Elbe one warm day, we need not be surprised anymore... Apparently, the immigration events of recent years in the southern North Sea represent only the tip of the iceberg that human action and inaction in climate protection has created on this planet. However, the corona virus pandemic in 2020 seems to have brought a temporary respite, as large parts of international air traffic in particular had to be shut down. This suggests, that flying is very likely to be much more harmful to the global climate than previously thought. It has long been known, that a single passenger aircraft produces as many emissions as about 10,000 cars. And yet the harmful gases are released promptly and without detour right into the uppermost layers of the earth's atmosphere. The result can now even be seen from space on a large scale: The deserts are getting bigger, the glaciers and freshwater reserves are dwindling, the Amazon-rainforest is burning, the Siberian Arctic is on fire, Australia is drier than ever and has to cope with huge bush fires. And the rainforests in Asia and Indonesia are also on the retreat. Parts of Europe are already transforming into steppes landscape. And in East Africa, droughts and plagues of locusts cause threatening famines such as the world had never seen before. If this continues, large parts of the planet will soon be hardly habitable. All this is no scaremongering; it is merely the quintessence of what the news delivers to our door every day. The Corona-pandemic of 2020 is a frightening reminder of how vulnerable and interdependent we have all become as a result of globalisation. But the things can`t continue as they are, even if the world's leaders deny them. It is easy to see how pathetic these people are by the fact, that they still only want to enrich themselves and hold on to power. Whether this planet goes to the dogs or not. However, there is an objective, higher truth and reality that neither can be locked up or ignored. And even if the **Fridays For Future** protests of a violent day were to fall silent for good, the subsequent heat waves and forest fires, as well as huge storms and floods, would still help our battered planet to regain its rightful place. And put all mankind back in the place where it naturally belongs. And then the point could also be reached very quickly, where it is only a question of the survival of the human species on this planet. And where greed and the pursuit of luxury and wealth become completely meaningless. In the southern North Sea, there is now a cut-throat competition between the most diverse organisms. The outcome is completely open and uncertain. If these developments continue, not only will a few fishermen on the German North Sea coast certainly go bankrupt, but chaotic environmental conditions could also make whole stretches of coast uninhabitable. And the collapse of mass tourism could then also result in very dramatic social upheavals and the emigration of hundreds of thousands. Therefore, investing in sustainable forms of energy and economy is a very good alternative to our current way of life. It is up to all of us to decide whether something changes. But if we do not change anything, Domina Natura will teach us some very bitter lessons, that we would have been better off without...

The following graphs are intended to provide a better illustration of the phenomenon currently visible in the southern North Sea. However, this is only a coarser view, not one that has been created by using scientific methods. It could, however, be regarded as a "working hypothesis" and then used as a basis for further considerations. Sometimes the species of marine life found do not necessarily say anything directly about the faunal changes, but rather by their size. For example, it was noticeable that all **Blackbelly rosefish (*Helicolenus dactylopterus*)** found between November 2019 and March 2020 were about 10 centimetres in size. This indicates that they all belong to the same generation. At the same time, finding this deep-water species in the shallow waters of tidal creeks, allows interesting conclusions to be drawn about how this species has already established itself in German waters. The same applies to the **Queen Scallop *Aequipecten opercularis.*** While this species was only rarely found in small numbers until 2018, catches of it off the East Frisian Islands increased in 2019, both in terms of quantity and size. This allows conclusions to be drawn about the growth periods and food supply of these mussels. This is similar to the **Otter Shell (*Lutraria lutraria*)**. Based on all these occurrences and other species belonging to subtropical faunal communities, it is easy to conclude, that the Mediterranean faunal fraction in the North Sea has already shifted at least 100 kilometres into the north. This means that some species, such as the **Tompot Blenny *Parablennius gattorugine*** or the **Dwarf Hermit *Diogenes pugilator*** did now permanently establish themselves in the southern North Sea. This means that they no longer migrate to the south during the winter months, but rather spend the winter in the deeper waters off the German islands. This is especially true when the temperature here remains about 10 to 11° Celsius(!) as in winter 2019/2020 and the winter season has obviously become "absent" in this sea area. These are no longer hypotheses, but facts, which can be objectively proven by the now permanent presence of the once southern marine fauna and the measurements of the oceanologists.

The course of the faunal boundaries in 1990 is based on the author's assessments based on various discussions with fishermen, various coastal inhabitants and marine biologists. As well as from other sources such as books, television documentaries and own observations.

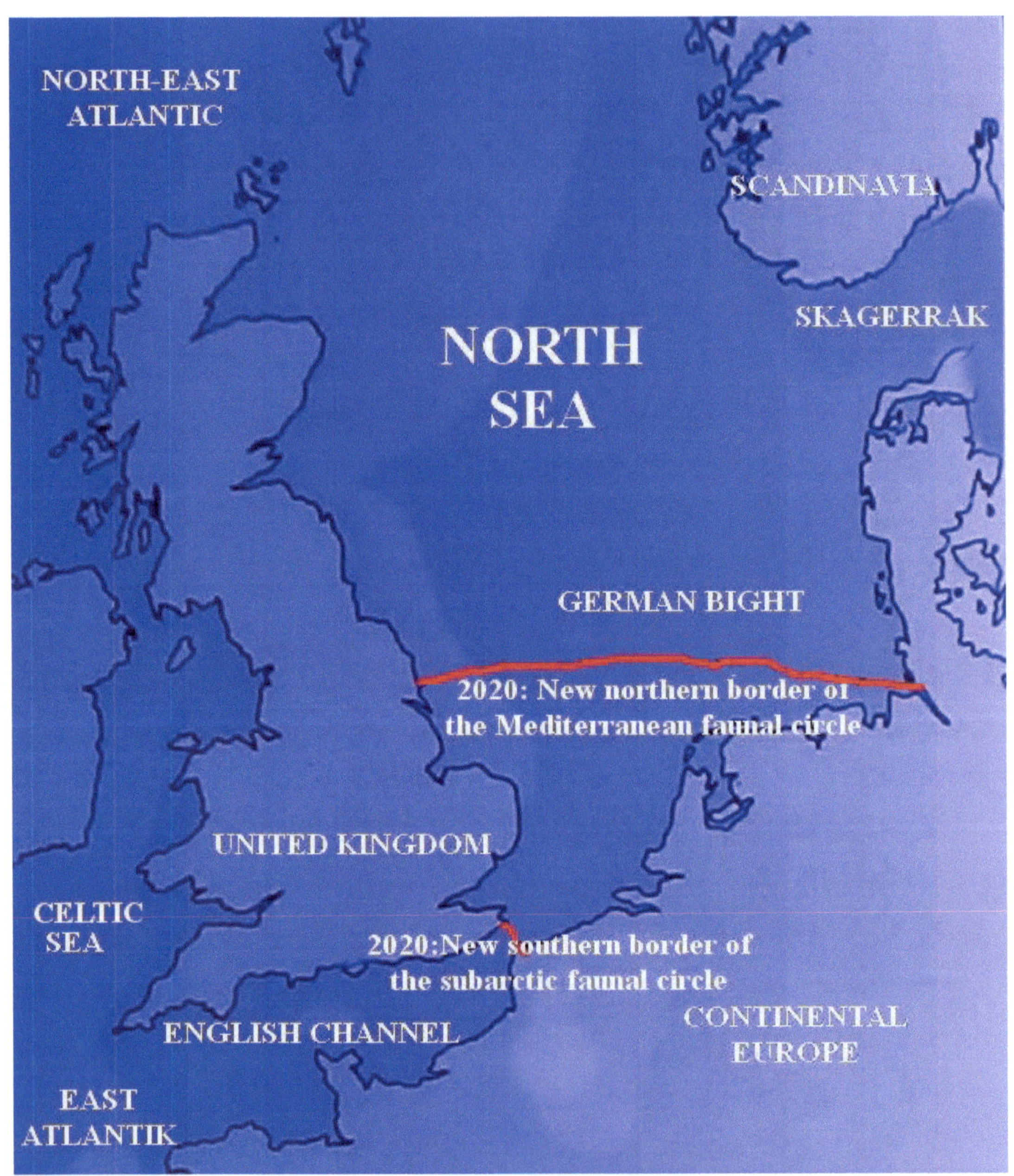

Faunal boundaries are not a static structure. However, they provide a rough guide to understanding the migration of species. The essence of the author's research is, that Arctic species such as the cod retreat northwards, while species such as the red mullet migrate northwards.

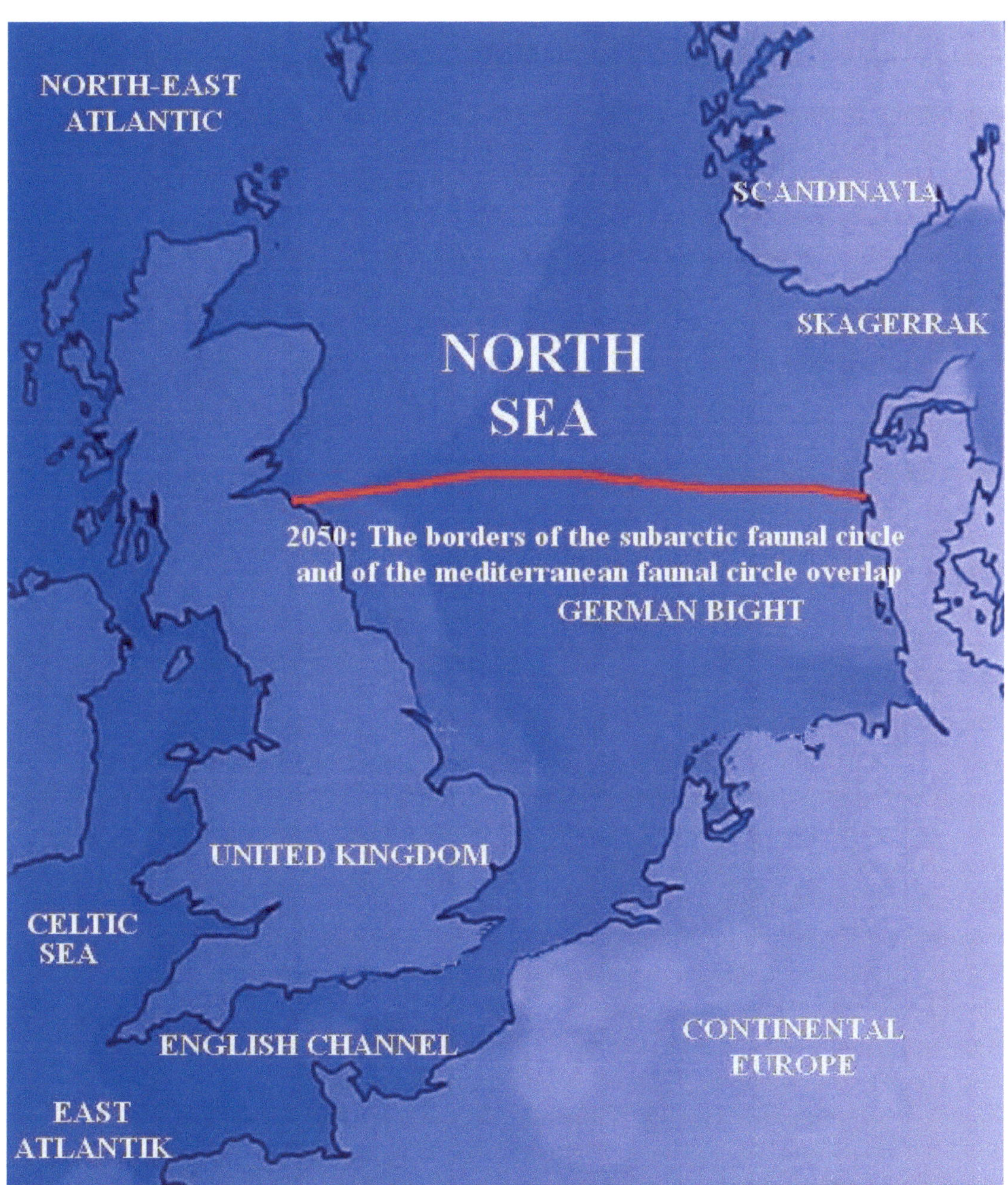

This is what it could look like in the North Sea in 30 years: The English Channel fauna did permanently establish in the German Bight. While subarctic species have almost completely left the German Bight. And the climate protection must now finally be considered a failure...

Atlanticum Bremerhaven, Forum Fischbahnhof, Schaufenster 6, 27572 Bremerhaven. Tel.: 0471-93233-0. E-Mail: Mail@forum-fishbahnhof.de; Domain: www.atlanticum.de.

Nationalpark-Haus Baltrum, Haus Nr. 177, 26579 Baltrum. Tel.: 04939-469. E-Mail: nlpe.baltrum@gmx.de.

Büsumer Meereswelten, Am Südstrand 9 A, 25761 Büsum. Inhaber: Gerhard Gebauer, Tel.: 0173-8625377. E-Mail: info@buesumer-meereswelten.de; Domain: www.buesumer-meereswelten.de.

Ostseestation Priwall, Priwallpromenade 29-31, 23570 Lübeck-Travemünde. Inhaber: Thorsten Walter GBR, Tel.: 04502-308705. E-Mail: info@ostseestation.de; Domain: WWW.Ostseestation-travemuende.de.

Aquarium Kiel, Düsternbrooker Weg 20, 24105 Kiel. Tel.: 0431-6001637. E-Mail: kontakt@aquarium-kiel.de; Domain: www.aquarium-kiel.de.

Sylt Aquarium, Gaadt 33, 25980 Westerland. Tel.: 04651-8362522. E-Mail: info@syltaquarium.de; Domain: www.syltaquarium.de.

Deutsches Museum für Meereskunde und Fischerei, Ozeaneum, Katharinenberg 14-20, 18439 Stralsund. Tel.: 03831-2650601. E-Mail: info@ozeaneum.de; Domain: www.ozeaneum.de

Multimar Wattforum Tönning, Am Robbenberg , 25832 Tönning, Tel.: 04861-9620-0. E-Mail: info@multimar-wattforum.de; Domain: www.multimar-wattforum.de.

Seehundstation Nationalparkhaus Norden-Norddeich, Dörper Weg 24, 26506 Norden. Tel.: 04931 - 89 19; das daran angeschlossene **Waloseum** befindet sich im Osterlooger Weg in Norden. E-Mail: info@seehundstation-norddeich.de; Domain: seehundstation-norddeich.de

Niedersächsisches Landesmuseum Hannover, Willy-Brandt-Allee 5, 30169 Hannover. Tel.: 0511-9807686. E-Mail: info@nlm-h.niedersachsen.de; Domain: landesmuseum-hannover.niedersachsen.de

Zoo-Aquarium Berlin, Budapester Str. 32, 10787 Berlin. Tel.: 030-254010. E-Mail: info@zoo-berlin.de; Domain: www.aquarium-berlin.de

Zoologisch-Botanischer Garten Wilhelma, Neckartalstr. , 70376 Stuttgart. Tel.: 0711-5402-0. E-Mail: info@wilhelma.de; Domain: www.wilhelma.de

Aquarium im Ostsee-Informations-Zentrum, Schiffbrücke 20, 24340 Eckernförde. Tel.: 04351 726266.

Haus der Natur, Museumsplatz 5, A-5020 Salzburg. Domain: www.hausdernatur.at

Haus des Meeres Vivarium, Fritz-Grünbaum-Platz 1, 1060 Wien (Mariahilf). Domain: www.haus-des-meeres.at.

Thanks!

I would like to thank all my friends and acquaintances, who have given me valuable tips for the creation of this work. Another thank you goes to the team of the Multimar-Wattforum in Tönning, who gave me many insights into the underwater world of the northern North Sea, to the team of the Nationalparkhaus Baltrum, and to the many hobbyists with whom I have exchanged experiences and sometimes animals over the years. Furthermore, I would like to thank the marine biologist Thorsten Walter from the Baltic Sea Station in Lübeck-Priwall for providing me with various pictures. Also I thank Stefanie Hamm for her beautiful detail photos. I am also grateful to various fishermen, national park rangers, employees of public institutions and aquariums, who gave me insight into the handling with the animals again and again. Without this rich exchange of information, this book in this form would hardly have come about.

Sven Gehrmann, in summer 2020.

Pictures:

Pictures by Stefanie Hamm:
Sepia officinalis(2X), 36; Alitta virens, 50; Chamäleon-Garnele(2X), 58; Processa canaliculata, 59.

Pictures by Thorsten Walter, Ostseestation:
Calanus finmarchicus, 54.

The copyright of these pictures remains with Thorsten Walter and Stefanie Hamm.

The reprint was made with friendly permission.

Literature and sources

Internet:

http://www.habitas.org.uk/marinelife/index.html
http://www.seawater.no/fauna/index.htm
http://www.fishbase.org/search.php
http://www.marinespecies.org
http://www.multimar-wattforum.de/
http://www.wikipedia.org/
http://www.wwf.de/themen/meere-kuesten/
http://www.glaucus.org.uk/Forum99.htm
http://zipcodezoo.com
http://www.marlin.ac.uk/species
http://www.vliz.be/Vmdcdata/macrobel/index.php
http://www.greenpeace.de
http://umweltanalytik.com
https://taz.de/Rekorde-in-Nord--und-Ostsee/!5023748/
https://www.spektrum.de/magazin/extrem-alte-muscheln/1435413
https://www.bsh.de
https://www.awi.de
https://www.br.de/klimawandel/klimawandel-kabeljau-dorsch-nordsee-sardinen-austern-100.html
http://taxonomicon.taxonomy.nl/
https://www.gmx.net/

Books and further literature:

Wilhelm Eigener, Enzyklopädie der Tiere, Georg Westermann Verlag, 1979, ISBN 3-14-508000-8.
Koie, Christiansen, Weitemeyer, Der große Kosmos Strandführer, Franck Kosmos Verlags GmbH, 2001, ISBN 3-440-08576-7.
Andrew C. Campbell, Der Kosmos Strandführer, Franckh`sche Verlagshandlung W. Keller & Co., Stuttgart. ISBN 3-440-04355-X.
Werner de Haas & Fredy Knorr, Was lebt im Meer an Europas Küsten? Stückle Druck & Verlag, 77955 Ettenheim. ISBN: 3-275-01302-5.
Georg Quedens, Strand und Wattenmeer, BLV Verlagsgesellschaft, ISBN 3-405-3805-1.
Klaus Janke & Bruno P. Kremer, Düne, Strand & Wattenmeer, Kosmos Verlags GmbH & Co., Stuttgart. ISBN: 3-440-09576-2.
Muus & Nielsen, Die Meeresfische Europas, Franck Kosmos Verlags GmbH & Co., Stuttgart. ISBN: 3-440-07804-3.
Gerd Pucka, Lehrbuch der Tierpräparation, Venatus Verlags GmbH, ISBN 3-932848-24-1.
Alwyne Wheeler, Das Grosse Buch der Fische, Verlag Eugen Ulmer Stuttgart. ISBN: 3-800170299.
Jörgen Möller Christensen, Die Fische der Nordsee, Franckh`sche Verlagshandlung W. Keller & Co. Stuttgart. ISBN: 3-440-04458-0.

Frank Emil Moen & Erling Svensen, Marine fish & invertebrates of Northern Europe, KOM 2004, ISBN: 0-9544060-2-8.

Paul Naylor, Great British Marine Animals, 2nd Edition, Sound Diving Publications, ISBN: 0952283158.

Alan Weisman, Die Welt ohne uns, Reise über eine unbevölkerte Erde, Piper Verlag, ISBN: 978-3-492-05132-3.

Elizabeth Kolbert, Das 6. Sterben, Wie der Mensch Naturgeschichte schreibt, Suhrkamp Verlag, ISBN: 978-3-518-42481-0.

J.L. Lozan, W. Lenz, E. Rachor, B. Watermann und H. v. Westernhagen, Warnsignale aus der Nordsee, Verlag Paul Parey, ISBN: 3-489-64634-7.

J.L. Lozan, W. Lenz, E. Rachor, K. Reise und H. v. Westernhagen, Warnsignale aus dem Wattenmeer, Blackwell Wissenschafts-Verlag Berlin, ISBN: 3-8263-3025-0.

Paul Sterry and Andrew Cleave, COLLINS COMPLETE GUIDE TO BRITISH COASTAL WILDLIFE, Collins Verlag, ISBN: 978-0-00-741385-0.

Markus Mahl, Meerwasser-Aquarium, Aquarium bauen und pflegen wie die Profis, Aquarium West GmbH, ISBN: 978-3-9819211-0-6.

Julian Cremona, Seashores, An Ecological Guide, Crowood Press, ISBN: 978-1-84797-804-2.

Wattenmeer, diverse Autoren, Wachholtz Verlag, ISBN: 3 529 05304 X (2. Auflage 1977)

Jose Tola, Die faszinierende Welt der Ökologie, Bassermann-Verlag, ISBN: 3-8094-0037-8.

John Pernetta, Großer Atlas der Meere, Naumann & Göbel, ISBN: 3-625-10746-5.

Christian Sardet, Plankton, Wonders Of The Drifting World, The University Of Chicago Press, ISBN-13: 978-0-226-18871-3; ISBN-10: 0-226-18871-X.

Malcolm Mac Garvin, Das Greenpeace-Buch der Nordsee, Franckh-Kosmos GmbH & Co., ISBN: 3-440-06207-4.

Patrick Louisy, Meeresfische Westeuropa Mittelmeer, Ulmer Verlag, ISBN: 3-8001-3844-1.

Armin Maywald, Wattenmeer, Im Wechsel der Gezeiten, Tecklenborg Verlag, ISBN: 3-924044-44-9.

Herbert Frei, Kathrin Herzer und Dr. Dieter Schmidt, Giftige und gefährliche Meerestiere, Müller Rüschlikon, ISBN: 978-3-275-01601-3.

About the author:

Sven Gehrmann, born in 1969, a native Berliner, who currently lives in Norden near Norddeich on the coast of Lower Saxony, has been involved with everything that lives under water since childhood. He has always been particularly interested and fascinated by crustaceans and fish. Since 1983 he is an enthusiastic hobby aquarist and nature fan of our native aquatic animals, especially the North Sea animals. So far, he has published various articles in aquaristic journals, ranging from North Sea animals to articles about anemone fish and various crustaceans. This was followed by publications on the fauna of the North Sea and the Mediterranean Sea, an eco-thriller and several volumes of poetry. In his publications he never minces his words and calls a spade a spade, as obviously nobody else does. In doing so, he takes no account whatsoever of a false sense of "political correctness", which has been successfully installed everywhere here to maintain the appearance of decency and freedom of opinion. On the other hand, he focuses on real "ecological correctness", even if it seems to be still in its infancy. And with it he makes himself unpopular to the one or the other institution. Sven Gehrmann is an avowed non-partisan democrat and sees himself as a neutral contemporary. In addition, he sees himself as a poet, thinker, philosopher, idealist and as a declared Protestant against today's world order which is hostile to nature, man and God. However, he distances himself from any form of fanaticism and sees himself as a friend of all those, who wish to make a constructive contribution to the preservation of our modern nature. Therefore, he also maintains contacts with alternative civil movements and is in constant dialogue with various interesting people from many spectrums of our society. His motto is: "Make friends and look what`s happening."

You can find him on the Internet at: **WWW.NORDSEEFAUNA.ORG.**

Actinia equina	41
Actinopterygii	104
Adna anglica	45,57
Aequipecten opercularis	27,162
Aequipecten opercularis radiata	27
Alitta virens	50
Ammodytes marinus	149
Annelida	47
Alosa fallax	108
Aphrodita aculeata	51
Arctica islandica	23
Asterias rubens	81,83,84
Atherina presbyter	112,113
Aurelia aurita	42
Austrominius modestus	56
Bathyraja brachyurops	101
Belone belone	106
Botrylloides leachii	92
Botryllus schlosseri	91
Buglossidium luteum	156
Calanus finmarchicus	54
Callionymus lyra	152
Cancer pagurus	79
Carcinus	18
Carcinus aestuarii	66
Carcinus maenas	65,79
Caryophyllia smithii	45,57
Cerianthus	41
Chelidonichthys cuculus	136
Chelidonichthys lucernus	136,137
Chelon labrosus	126
Chironex fleckeri	40
Chorda	89
Chrysaora hysoscella	43

Ciliata mustela	122
Ciona intestinalis	90
Cirripedia	56
Clupea harengus	109
Cnidaria	39
Coelenterata	39
Copepoda	54
Copula	78
Crangon crangon	70
Crepidula excavata	20
Crepidula fornicata	20
Crinoidea	81
Crustacea	52
Ctenolabrus rupestris	148
Ctenophora	39,46
Cumacea	55
Cyanea capillata	42
Cyanea lamarcki	42
Dasyatis pastinaca	95,103
Decapoda	58
Dentex maroccanus	141
Diadumene cincta	44
Diastylis rathkei	55
Dicentrarchus labrax	127
Diogenes pugilator	58,75,162
Echiichthys vipera	144
Echinodermata	81
Echinus esculentus	87
Elasmobranchii	94
Enchelyopus cimbrius	123
Engraulis encrasicola	112
Entelurus aequoreus	133
Errantia	47
Eutrigla gurnardus	138
Gadus morhua	118
Galathea strigosa	74

Galeorhinus galeus	99
Gasterosteus aculeatus	128
Gobius niger	150,151
Hediste diversicolor	48
Helicolenus dactylopterus	135,162
Hemigrapsus penicillatus	77
Hemigrapsus sanguineus	77
Henricia sanguinolenta	83
Hippocampus	130
Hippocampus hippocampus	130
Hippolyte varians	58
Homarus gammarus	79
Hyas araneus	78
Hydra	39
Illex spec.	35
Lamna nasus	95
Lepas anatifera	57
Liocarcinus marmoreus	70
Liocarcinus navigator	71
Loligo forbesii	32
Loligo todarus	35
Lophius piscatorius	159
Lota lota	121
Lutraria lutraria	30,162
Macropodia linaresi	68
Magallana gigas	18,24,44
Maja brachydactyla	69
Maja squinado	69
Marthasterias glacialis	88
Merlangius merlangus	117
Merluccius merluccius	124
Merluccius senegalensis	124
Metridium senile	44
Mimachlamys varia	27,28
Mnemiopsis leydyi	154
Mola mola	125

Mollusca	19
Mullus surmuletus	139
Mya arenaria	29,31
Mysis spp.	38
Mytilus	18
Mytilus edulis	24
Mytilus galloprovincialis	28
Necora puber	72
Neogobius melanostomus	151
Nephtys hombergii	49
Nereididae	47
Nereis	47
Ommastrephes sagittatus	35
Ommatostrephes sagittatus	35
Osmerus esperlanus	116
Ostreidae	22
Pagurus bernhardus	76
Palaemon	53,61,62
Palaemon adspersus	61,62
Palaemon elegans	61,63
Palaemon macrodactylus	60,62
Palaemon serratus	61,62
Pandalus montagui	53,64
Parablennius gattorugine	107,162
Pegusa lascaris	156
Perforatus perforatus	57
Pleurobrachia rhodopis	46
Physalia physalis	40
Platichthys flesus	155
Pleuronectes platessa	154
Pododesmus squama	22
Pollachius virens	119
Portumnus latipes	8,73
Processa canaliculata	59
Psammechinus miliaris	83,86
Raja clavata	101,102

Rhizostoma pulmo	43
Salmo salar	115
Salmo trutta trutta	114
Sarda sarda	146
Sardina pilchardus	109,111
Sarpa salpa	142
Scomber scombrus	147
Scorpaena scrofa	134
Scyliorhinus canicula	97
Scyphozoa	41
Sepia officinalis	36
Solea solea	157
Sparus aurata	143
Spinachia spinachia	129
Sprattus sprattus	109,110
Squalus acanthias	96
Styela clava	93
Syngnathus acus	131
Syngnathus rostellatus	132
Todarodes sagittatus	34
Torpedo marmorata	95,100
Trachinidae	145
Trachinus draco	144,145
Trachurus trachurus	140
Trisopterus luscus	120
Tunicata	89
Vibrio cholerae	17
Vibrio vulnificus	17
Xaiva biguttata	67
Xiphias gladius	153
Zeus faber	158

Bibliographic information of the German National Library:
The Deutsche Nationalbibliothek lists this publication in the German National Bibliography; detailed bibliographic data are available on the Internet at http://dnb.d-nb.de.

Imprint: Fire! In the Wadden Sea…
Or what the fauna of the North Sea wants to tell us very urgently

Production and publishing: BoD - Books on Demand, Norderstedt.
ISBN: 9783751980654.

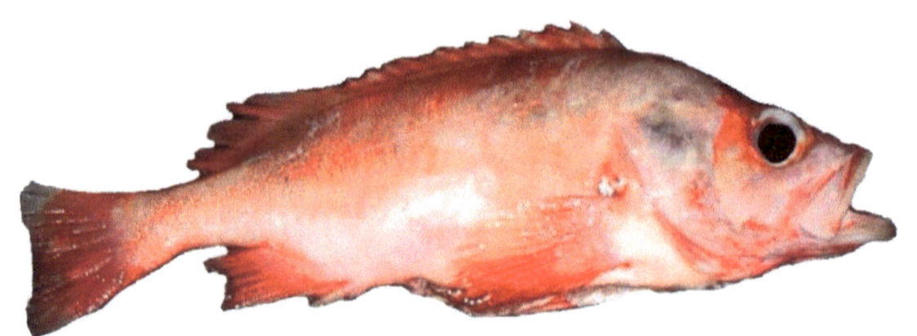